MONSTERS AMONG US

The Truth About a Hidden World You Never Knew

Andrea McGann Keech

Pulp Hero Press
The Most Dangerous Books on Earth
www.PulpHeroPress.com

Editor: Bob McLain
Layout: Artisanal Text

ISBN 978-1-68390-063-4
Printed in the United States of America

Pulp Hero Press | www.PulpHeroPress.com
Address queries to bob@pulpheropress.com

For Katherine,
the sweet and brave

CONTENTS

Introduction
A Real "Dinosaur" Fish vii

Chapter One
The Kraken 1

Chapter Two
The Beasts of Bodmin Moor and Exmoor 7

Chapter Three
The Sasquatch 13

Chapter Four
The Chupacabra 19

Chapter Five
The Loch Ness Monster 23

Chapter Six
The Werewolf 29

Chapter Seven
The Yeti 37

Chapter Eight
The Sea Serpent 43

Chapter Nine
The Gloucester Sea Serpent 51

Chapter Ten
The Globster 61

Chapter Eleven
Three Notorious Globsters:
Gambo, Trunko, and the Stronstay Beast 67

Chapter Twelve
The Beast of Gévaudan 71

Chapter Thirteen
The Thunderbird 79

Chapter Fourteen
The Kelpie 87

Chapter Fifteen
The Mermaid 95

Chapter Sixteen
The Mahamba, the Neguma-monene,
and the Mokele-mbembe 103

Chapter Seventeen
The Nandi Bear 109

Chapter Eighteen
The Tasmanian Tiger and the Queensland Tiger 115

Chapter Nineteen
The Megalodon 123

Chapter Twenty
The Mothman 129

Chapter Twenty-One
The Wendigo 137

Chapter Twenty-Two
Hype, Hokum, and Historical Hoaxes 145

Acknowledgments 153

About the Author 155

A Real "Dinosaur" Fish

Either we do know all the varieties of beings which people our planet, or we do not. If we do not know them all—if Nature has still secrets in the deeps for us, nothing is more conformable to reason than to admit the existence of fishes, or cetaceans of other kinds, or even of new species...which an accident of some sort has brought at long intervals to the upper level of the ocean.

— Jules Verne
Twenty Thousand Leagues Under the Sea (1870)

If you're like most people, you're already well acquainted with the many myths and legends about monsters who share our world. You know about Sasquatch, the reclusive Bigfoot creature who inhabits the deep, remote woods of the Pacific Northwest. He's been domesticated in the film *Harry and the Hendersons*. You're likewise familiar with the Himalayan Yeti, also known as the Abominable Snowman, who has been transformed from monster to marshmallow in a series of cartoons and popular children's storybooks like *The Yeti Files*. He even makes a guest appearance on roller-coaster attractions at Disneyland and Walt Disney World. Giant killer sharks, dragons of the land and sea, the Loch Ness monster, and a host of other fabulous, fanciful beasts are a part of our collective popular consciousness, a part that many sane people treat with healthy skepticism if not downright disbelief—but what if some of those legends and stories were true? In this book, we'll take a closer look at the most infamous of the world's monster tales and attempt to separate fact from fiction. You'll be surprised!

As you read these stories, I want you to keep an open mind. The most persuasive argument I can provide for doing so is the fact that a prehistoric marine creature known to have become extinct with the dinosaurs over 65.5 million years ago was discovered to be alive and well just seventy-eight years ago. Well, it wasn't alive, exactly, but it had just been hauled aboard a fishing boat. That's right, a "dinosaur" fish from the end of the Cretaceous period (the last and longest segment of the Mesozoic Era) showed up on a fishing vessel off the eastern coast of South Africa on December 22, 1938. Although it was caught by Captain Hendrick Goosen, it was thirty-one year old Marjorie Courtenay-Latimer (1907–2004) who recognized the significance of the mysterious fish. Having nothing more to go on but her familiarity with the fossil record, she recognized a prehistoric anomaly when she saw it. To honor Marjorie's perspicacity in recognizing the significance of this discovery, the species was given the scientific designation *Latimeria chalumnae* after her surname and the name of the river near where it was found, the Chalumna.

There is a designation for animals believed to have disappeared completely from the fossil record, only to reappear many years later. It's called the *Lazarus taxon*, referring to the Biblical account in the Gospel of John when Jesus raised Lazarus from the dead. There are scientifically well-documented, indisputable, widely accepted accounts of scores of discoveries like the coelacanth, but it is the strange and homely fish we'll consider first as a representative of this bizarre and unusual taxonomy.

Marjorie Courtney-Latimer was born prematurely and was ill for much of her young life. She kept herself busy with the study of birds and animals. An autodidact, and was hired at twenty-four as a curator by the East London Museum, Eastern Cape, South Africa. She let it be known that she wanted to be notified of any rare or unusual fish or sea life caught in the waters near where she lived. That is how she came to be at the right place at the right time to make her astonishing discovery:

> I picked away at the layers of slime to reveal the most beautiful fish I had ever seen. It was five feet long, a pale mauvy blue with faint flecks of whitish spots; it had an iridescent silver-blue-green sheen all over. It was covered in hard scales, and it had four limb-like fins and a strange puppy dog tail.

The young museum curator was unable to identify the fish in her books. The only way to preserve the specimen was by taxidermy. Once her friend, amateur ichthyologist and senior chemistry lecturer at Rhodes University, L.J.B. Smith saw it, he wrote: "There was not a shadow of a doubt. It could have been one of those creatures of 200 million years ago come alive again." He detailed the account in his book *Old Fourlegs: The Story of the Coelacanth* (Longmans Green, London, 1956). This line is particularly descriptive of the emotion he felt: 'A bomb seemed to burst in my brain'. As he would likely have put it today, the discovery was mind-blowing.

Coelacanths appeared on earth about 400 million years ago and were believed to have vanished entirely 340 million years after that. Seeing one in 1938, L.J.B. Smith observed, was "as if a dinosaur had suddenly been found alive and well and roaming the Earth."

Since Marjorie's original encounter, coelacanths have been found in the waters of the Comoro Islands, iSimangaliso Wetland Park, Kenya, KwaZulu-Natal, Mozambique, Madagascar, and Tanzania, but it wasn't until 1952 that another specimen finally resurfaced, this time in the Comoro Islands north of Madagascar. These fish remain in the ocean at depths of around six-hundred feet and don't live long in the shallows. About two hundred have been caught since they were discovered, and perhaps two hundred of the now-protected species are believed to survive today.

A *second* entirely unexpected, new species of coelacanth was discovered by UC Berkeley researcher Dr. Mark Erdmann in 1998 in Sulawesi, Indonesia. It was found 6,200 miles from the area where the original "dinosaur" fish was found. Local fishermen call it the *raja laut*, "king of the sea." Surprise, science! Not one but *two* types of living fossils surfaced.

Is the coelacanth really the *only* dinosaur-like creature still living among us? Amphibians, birds, fish, mammals, mollusks, plants, protostomes (worms, snails, mollusks, and arthropods), and reptiles all have *Lazarus taxon* individuals, once believed to be extinct, that have "miraculously" reappeared. Why, then, should it require such a stretch of the imagination to think discoveries like the coelacanth are somehow unique, that such a thing can't possibly happen again?

Our planet still harbors locations so inhospitably remote and inaccessible that remnants of vanished species might have managed to retain small populations of survivors. Mastodons, after all, remained viable on tiny Wrangle Island off Siberia long after their mass extinction on the mainland. Ten thousand years ago, most of the mastodons were gone. As recently as four thousand years ago, however, this little "pocket population" lived on in splendid isolation.

Not until 1824, when Mary Anning found the physical evidence, did people believe that dinosaurs had ever existed on Earth before us. She made a modest living finding and selling marine fossils at Lyme Regis along the English Channel, including Jurassic fossils of the ichthyosaur, plesiosaur, and pterosaur (then called the "flying dragon"). If you've ever heard this 1908 tongue twister (and who hasn't?), "She sells sea shells by the sea shore," it was an homage to Mary. Fewer than two hundred years ago, humans believed the animal world started with us.

The gorilla was a shocking sensation in Victorian times. In 1847, Thomas Savage obtained a great ape's skull and bones, but that wasn't enough to convince a skeptical public. Only in 1859, when Paul Du Chaillu exhibited some twenty skins of gorillas he had killed in Gabon over a four-year period, was credence finally given to the legend of the hairy men called *njena* by the locals. Edward Stanton taunted Abraham Lincoln with the unkind epithet "the Original Gorilla" in 1861 "when that slur was shockingly fresh," as David Quammen writes in an April 4, 2013, *New York Times* book review of Monte Reel's *Between Man and Beast: An Unlikely Explorer, the Evolution Debates, and the African Adventure That Took the Victorian World by Storm* (Doubleday, 2013). In the fascinating book, you'll see how easily and irrevocably what the world believed about itself was turned on its ear by a single fantastic discovery.

Similarly, the giant panda until 1869, the giant squid until 1861, the Komodo dragon until 1910, the manatee until 1493, the platypus until 1799, and the okapi until 1901 were each thought to be mythological creatures of legend as opposed to real, living animals. Even when presented with dead specimens, scientists doubted the evidence they saw with their own eyes. English zoologist George Shaw suspected the body of a platypus he examined might just be a hoax: "On a subject so extraordinary as the present, a degree of skepticism is not only pardonable, but laudable; and I ought perhaps to acknowledge that I almost doubt the testimony of my own eyes."

Christopher Columbus wrote about his encounter with West Indian manatees in his ship's log, believing they were mermaids, albeit exceptionally ugly ones with "masculine features." No wonder people were skeptical of such reports, especially after P.T. Barnum's phenomenally popular American Museum exhibit of his Fiji Mermaid turned out to be nothing more than the torso and head of a monkey sewn to the body and tail of a fish!

Several examples of mythological creatures turned out to be a very real species of animal. If it has happened before, happened over and over again through the centuries, why are we still so reluctant to concede that it quite possibly might happen again? That's the premise of this book. We'll take a look at the familiar stories of fantastic creatures and their plausibility. Could monsters be living among us largely undetected? Let's see.

CHAPTER ONE

The Kraken

An impossibly large, multi-limbed, graceful creature drifts fathoms below the little fishing boat in the ice-cold waters of the Norwegian Sea. It trolls for prey more than three-thousand feet down in the inky blackness, ravenous with hunger. It would eat its own kind if it came upon a smaller individual, but no such luck presents itself this day.

The fisherman's nets are likewise as empty as the gut of the beast. He heaves a deep sigh of despair, for without a decent catch of fish, his family will go hungry. No food comes, not to the squid nor to the man. Slowly, the siphon jet the tremendous cephalopod uses to propel itself points it in desperation toward the surface. Something is there, something that might satisfy its searing need to feed. Two eyes the size of dinner plates attempt to focus in the unaccustomed pale light that filters through the choppy surface of the sea.

It is uncannily clever, able to solve problems and to reason. The silhouette of the man is intriguing, and it has run out of options. Slowly, deliberately, one of its two long feeding tentacles snakes above the gray surface and wraps itself gently around the man's waist in a firm caress. Too late, the man flails about in a futile attempt to wrap his hands around an oar, a gaff, anything. In an instant, he has gone from predator to prey. Pulled into the icy darkness below, his last conscious thought is of the shape of the underbelly of his boat as it rapidly disappears above him.

It will eat slowly, deliberately, savoring each bite ripped from the soft body with its sharp beak. Who can tell when the next opportunity may come?

The late thirteenth-century saga of Iceland, the Örvar-Oddr, makes reference to a beast widely considered to be the Kraken:

"The hafgufa is the hugest monster in the sea. It is the nature of this creature to swallow men and ships, and even whales and everything else within reach." Today, when someone says, "Release the Kraken!" it's akin to Shakespeare's line, "Cry 'Havoc!' and let slip the dogs of war." The fictional order was voiced by Zeus in *Clash of the Titans* (1981) and is now most often delivered tongue-in-cheek, in the same way a modern brand of beer and a violent video game are named Kraken, but making light of a deadly attack by the beast wasn't always the case.

An anonymous Norse author wrote in *Konungs skuggsjá* about 1250: "There is a fish that is still unmentioned, which it is scarcely advisable to speak about on account of its size, because it will seem to most people incredible." Swede Carl Linnaeus in 1735, known as the father of modern taxonomy, classified the Kraken as a cephalopod (the word literally means "head-foot") and called it a "unique monster," although he admitted to not having seen it himself. Danish Bishop Erik Pontoppidan, writing in *The Natural History of Norway* (Copenhagen, 1752), claimed that if the arms of the Kraken were to grasp hold of the largest man-of-war, they could pull the ship down to the bottom of the ocean. Finally, the Swedish author Jacob Wallenberg in 1781 wrote about the Kraken in a work titled *My Son on the Galley*. He said, "Kraken, also called the Crab-fish, which is not that huge, for heads and tails counted, he is no larger than our Öland is wide (slightly less than ten miles)." That's still one *very* big Kraken, as Öland is the second largest of the Swedish islands and inhabited by about twenty-five thousand people!

Sailors, mostly from Norway and Greenland, encountered enormous beasts in the frigid northern seas and tried to describe them to the best of their abilities. The first accounts didn't mention tentacles, a feature that is now accepted as a requisite part of Kraken anatomy. The animals were first described more like giant crabs or even whales. A gorgeous drawing done in 1801 by Pierre Dénys de Montfort depicts an enormous octopus with its tentacles wrapped around the masts of a large schooner ready to pull it under, crew and all. Its eyes are like giant saucers. An octopus has eyes very similar to those of humans, but the eyes of the giant squid are the largest of any creature on earth. Those round, staring orbs de Montfort drew look not like the eyes of an octopus but very much like the eyes of a giant squid. Danish naturalist Japetus Steenstrup identified a large squid's beak that had washed ashore in Denmark in 1857. It was, he concluded, physical proof of the Kraken legend. It was from an animal he dubbed *Architeuthis dux,* the ruling squid.

In *Moby-Dick,* 1851, Herman Melville wrote about Captain Ahab's whaling ship, the *Pequod,* coming upon "great live squid, which, they say, few whale-ships ever beheld, and returned to their ports to tell of it." This chapter ends by saying the giant squid "is included among the class of cuttle-fish, to which, indeed, in certain external respects it would seem to belong, but only as the Anak [Biblical founder] of the tribe." The giant squid in Jules Verne's *Twenty Thousand Leagues Under the Sea,* 1870, continued to embellish the fierce reputation of this legendary beast.

James Owen writing for *National Geographic News* in an article published on April 23, 2003, confirmed an encounter in New Zealand as the first live sighting of a colossal squid. It was dubbed *Mesonychoteuthis hamiltoni* by scientists who examined the squid. They suggested calling it "colossal squid" to distinguish it from the giant squid (*Architeuthis dux*). The species was described as "the biggest and most fearsome squid known to science" and one that could grow to 40 feet.

A colossal squid found in the Ross Sea is the second complete animal ever found. This cephalopod had a proportionately large beak and hooks along its tentacles. It's only natural for a discovery like this one to spark strong interest in the old Kraken legends.

Steve O'Shea, Auckland University of Technology, called it "a true monster," and that description seems fair. A monster is an imaginary creature that is large, ugly, and frightening. *Mesony-choteuthis hamiltoni* is all of these—except imaginary. O'Shea told the BBC: "Giant squid is no longer the largest squid that's out there. We've got something that's even larger, and not just larger but an order of magnitude meaner," research associate Kat Bolstad, research assistant at Auckland University of Technology, observed in the same interview. "This animal, armed as it is with the hooks and the beak that it has, not only is colossal in size but is going to be a phenomenal predator and something you are not going to want to meet in the water."

One of my favorite writers on the topic of marine life, Richard Ellis, is a research associate at the American Museum of Natural History. He isn't given to hyperbole, and has a cool head with a tremendous amount of perspective. Ellis points out that the colossal squid "is no more a monster than *Architeuthis* is." He says, "I wrote *The Search for the Giant Squid* (Penguin Books, 1998) to try and dispel some of the crazy ideas that this cephalopod is in any way dangerous to humans, and the same holds true for *Mesonychoteuthis*." It's an excellent book and one I highly recommend, along with his well-known books on sharks, like *The Great White Shark* (Stanford University Press, 1991). Even if the colossal squid is not dangerous, people at sea who come upon a tremendous creature of these dimensions might be forgiven for being at least somewhat concerned for their safety.

Not until 2004 was a living squid filmed in its deep, natural habitat. Specimens found at the surface are often gravely ill or nearly dead. Tsunemi Kubodera of the National Science Museum in Tokyo, Japan, joined forces with whale expert Kyoichi Mori. The two men used the known locations of sperm whales and took a still photo of a live giant squid near the Ogasawara Islands in the North Pacific, the very first such photo. In 2012, Kubodera took the first video footage of a giant squid. It took four hundred hours in a small sub to secure it, but the film was definitive proof that the giant squid is, indeed, a predator and not simply a scavenger.

At Mount Holyoke College in South Hadley, Massachusetts, Mark McMenamin made a convincing argument in 2011 that during the age of the dinosaurs, there may well have been outsized squids as long as a hundred feet or more. (The largest modern squid identified, by comparison, was forty-three feet long and weighed close to a ton.) These prehistoric Krakens may have preyed on ichthyosaurs, giant marine reptiles that looked a bit like modern dolphins. (You will recall that it was the discovery of an ichthyosaur skeleton that propelled Mary Anning to fame.) Ichthyosaur vertebrae were found to be arranged in a pattern that closely resembles the discs on a cephalopod's tentacles. McMenamin's hypothesis is that a kraken took the marine reptiles back to its lair to consume, thus arranging the bones in the characteristic pattern. He has also found a fossil of what he believes to be the giant beak of the prehistoric beast.

Assuming that ancient tales of the Kraken were attempts to describe the little-known, misunderstood giant squid or possibly the colossal squid, then who is to say how large some of them may grow to be? The first motion picture dates back to 1878, but it took two hundred and thirty-four years after that to finally capture a moving image of the greatest of the squids in its own habitat. There is so little we know for certain about the creatures of the deepest ocean regions. At mostly uncharted depths of up to 36,200 feet, it isn't out of the realm of the possible to imagine a Kraken of one hundred feet or so surviving silently in the dark as it always has, by stealth and cunning.

Think about that next time you toss back a cold Kraken beer (with its slogan, "Put a beast in your belly!") while playing *Kraken Battle: Part 1* on your Xbox 360.

CHAPTER TWO

The Beasts of Bodmin Moor and Exmoor

The heartwarming scene is one occurring repeatedly throughout all of England in the springtime. It's lambing season, and the newest member of the flock has just raised itself up on wobbly and unsteady baby legs. A proud mother licks it clean and quickly nudges it forward in the deepening twilight, encouraging it to walk. It *must* walk. Eagerly, repeatedly, it butts the mother's side with its fuzzy head and nuzzles to find the source of life, her warm, rich milk.

Instinct tells the mother that this is the most vulnerable period in her offspring's young life. Millennia of predation by wolves, mountain lions, coyotes, and canines have imprinted upon her consciousness the immediate need to get the baby up and away from the place of its birth as soon as possible in order to ensure its survival. This place smells of blood, and that scent will swiftly attract any predators lurking nearby.

Another baby, though, is also hungry. It waits patiently with four siblings for its mother's return. It keeps still and cuddles closely with its fellows for warmth in a den deep within the dark shaft of an abandoned mine on the windswept, bleak moors of Cornwall. A large, tawny cat crouches motionless in the shadows, watching the baby lamb take its first steps. The smell that means food, the insistent bleating of the newborn lamb, and the anxious sounds made by the mother sheep have attracted its notice. Yellow eyes glint in anticipation, but patience is her stock in trade. She won't attack until the moment is exactly right.

Wide eyes shine florescent yellow-green, caught in a shaft of moonlight from the rising of the big yellow circle in the sky. An upper lip curls to reveal a row of needle-like fangs that resemble a row of little daggers with four sharp, elongated cuspids. Suddenly, a snarl escapes from her throat, and her scream rips apart the quiet night. She pounces, secure in the knowledge that this will be an easy kill. Her kittens are safe, and they will be eagerly awaiting her return.

The Beast of Bodmin Moor was first sighted in 1983 and has been seen some sixty times since then. Imagine a map of England. On the western side at the bottom is the big toe of the Cornwall peninsula extending into the chilly Atlantic Ocean. In the center of Cornwall is the large, remote moor called Bodmin. Six months of government investigation and close to a quarter-of-a-million dollars spent in the effort produced no credible evidence but did rule out briefly intriguing photos of what proved to be a twelve-inch, black house cat. Still, the sightings persisted. A week after the government issued its conclusion in 1995, a teenaged boy discovered the skull of an extremely large cat in the River Foley area. Interest rose, along with strong suspicion of government incompetence. The Natural History Museum in London determined that it was the head from a leopard skin rug, a head neatly sawn from the backbone with the skull evidencing scraping consistent with taxidermy.

The *BBC News Online*, on July 28, 1998, posted a twenty-second-long video of what some thought was surely the Beast of Bodmin, finally captured on tape. It was supposed to show a black, three-and-a-half foot long feline. In 2014, this footage was determined to show—yet again— nothing more than a black house cat. The director of the Newquay Zoo and big cat expert, Mike Thomas, believed at the time it was taken that the video was "the best evidence yet" of big cats roaming Bodmin Moor, he told the BBC.

While that evidence didn't pan out, Thomas' second idea may well have merit. The BBC further reported, "He believes the animal could be a species of wild cat which was supposed to have become extinct in Britain more than a century ago." Thomas is referring to the Scottish wildcat, *Felis silvestris*, a fierce predator that once ranged south into Wales and England but whose population is now limited to some forty individuals clinging to viability in the most remote northern reaches of Scotland. They are rarer than tigers. Could a few breeding pairs have sought refuge in the abandoned mines and desolate moors of Corn-wall a hundred years ago instead of moving north? This, the last native cat species of Britain, may be on the way to extinction, yet small scattered populations could certainly persist after the majority of the Scottish wildcats were gone from their former range in England.

The Beast of Exmoor has much in common with that of Bodmin Moor. An English farmer from South Moulton, Eric Ley, lost over one hundred sheep in a three-month span in 1983. Sightings of black or tan big cats resembling a mountain lion or a black leopard ranging in size from four to eight feet (measured nose to tail) have been recorded. Such reports were taken seri-ously enough by the Ministry of Agriculture that Royal Marine snipers were sent into the area to shoot the Beast of Exmoor. The captain of the division also reported sightings by the Marines and said that the big cat "always moved with surrounding cover amongst hedges and woods." After the snipers departed, the attacks increased again.

The *Lazarus taxon* is a way of explaining the fact that a few remaining individuals of the Scottish wildcat species, a native species relatively recently extinct or very nearly so, could be what residents have been seeing and reporting over the years. The description of the animal certainly fits, as does the behavior. There are so many reports over long spans of time that it seems reasonable to suspect *something* is very likely prowling around the countryside. There is also another plausible hypothesis.

In Great Britain, it was perfectly legal and quite fashion-forward to keep large, exotic cats as companion pets. Lion cubs were even sold at Harrods department store! Christian, an African lion, was purchased by John Rendall and Anthony Bourke in 1969 at a cost of 1,500 pounds, about $4,600 in today's dollars. Their cub rode along with them in their convertible and was a favorite among the tony clients at their upscale furniture store appropriately called Sophisticat on King's Road. A strong bond was forged between the two men and the small lion, but keeping him fed and happy became increasingly difficult as he grew to adult size. Assistance from George Adamson, of *Born Free* fame, was sought. Adamson managed to successfully reintroduce Christian into the wild.

A year later, the now nearly mature lion's former owners sought him out. Would he remember them? The 1971 reunion was nothing short of joyful. Not only did Christian remember his beloved friends, he introduced them to his two lady lion paramours, who also graciously accepted their company. A film of the meeting received more than 18 million views on YouTube. Again in 1973, Rendall and Bourke staged a reunion, their last. Christian was twice the size he had been when released, had a pride and cubs of his own, and approached them stiffly and slowly. Once he had satisfactorily confirmed their identity, though, he rushed at them like a playful kitten, hugged them around the shoulders just as he had always done, and generally behaved like a friendly house cat. Christian wasn't seen again after that meeting, and it was assumed he had died.

It was certainly an interesting period in the UK when Christian lived there. Well-heeled ladies could be seen strolling about greater London with ocelots, lynxes, cheetahs, or even leopards on leashes like swanky urban Cleopatras. Owners of the exotic cats suffered no sanctions or consequences whatsoever. In America at the same time, such things were also reasonably common. A friend of mine kept a pet ocelot in a cage in her backyard. Adjacent to my school bus stop, a cougar was kept caged behind a neighbor's garage. You may recall Katherine Hepburn and Cary Grant with a tame leopard in the 1938 madcap comedy *Bringing Up Baby*. In 1976, however, the British government passed the Dangerous Wild Animals Act to protect the public and ensure the animals' welfare. After the law was passed in the United Kingdom, owners of these big, beautiful beasts had few options.

First, they could seek a license to keep their big cats, a considerably costly, daunting, and tangled-in-red-tape bureaucratic process. Second, they could give their pet to a licensed zoo or licensed private individual willing to accept it. As you might imagine, such places would be difficult to secure given the large number of pets seeking sanctuary. Finally, they could have their pet euthanized. But wait! A fourth possibility, of course, springs immediately to mind, doesn't it? They could simply release their much-loved animal friends into a remote wilderness area and hope their former pets could manage to survive by preying upon the local game. Which option would you have chosen? Me, too.

At precisely the time it was no longer legal to keep exotic pets, stories of large cats began popping up in local news accounts across Great Britain. There have been more than fifty reports in northern Wales alone. Livestock was found mauled and partially eaten. Deer carcasses turned up consumed in the characteristic way large cats eat their prey. There was established a British Big Cats Society dedicated to the study of these animals; it maintains a detailed map of sightings.

On August 26, 2000, Josh Hopkins, eleven years old, was clawed across the face in Trelleck, in southern Wales, by what was thought to be a lynx. On May 4, 2001, a female European Lynx was discovered in Cricklewood, North London, after being spotted in someone's backyard. It was injured and taken to London Zoo. The lynx, in fact, was formerly a native species in England, now believed to have been wiped out. Fiona Keating in the *National Business Times* (January 2016) reports that plans are afoot to reintroduce breeding pairs of lynxes into the same forests where they dwelled some 1,300 years ago.

Just once did I get up close and personal with a big cat, a black panther, and I'll never forget it. It was early morning at the city zoo in Portland, Oregon, where I attended what was innocently described as "Breakfast with the Beasts." We were duly given breakfast accompanied by a lecture about the various big cats in residence at the zoo. Then, we were treated to the sight of them being fed their own breakfast which consisted of large, bloody chunks of meat. Backstage, away from the public area of the zoo, we walked a narrow channel between two banks of cages filled with big cats. There was perhaps a foot of clearance on each side of the path.

There came a high-pitched scream that ended in a fierce snarl, incredibly loud and right beside where I stood. Quicker than you can say "fight or flight," the panther rushed the fence. My heart was in my throat, and if you don't know that oft-described sensation, it's because you, my friend, have never been treated to "Breakfast with the Beasts." Dagger-like white fangs were bared, yellow eyes narrowed to slits, ears laid back tight against its head, every black-on-black spot was visible on the panther's gleaming pelt. The visceral sense of terror was immediate. That memorable experience gave me an extremely healthy, lasting respect for big cats.

The biggest of the felines are naturally nocturnal and mostly shy of encounters with humans. They generally prefer to avoid people whenever possible, which is fortunate for us. Still the "scepter'd isle" isn't very big. It's entirely likely that the big cats of Britain are real and that they are the descendants of those creatures released by their owners in 1976 as well as the last remaining vestiges of the native Scottish wildcats who have managed to survive largely undetected in inhospitable circumstances. Reports still continue to come in of big cats roaming the moors of Bodmin and of Exmoor. Livestock disappears or is mutilated. Game animals like deer are being targeted and eaten by *something*. As things stand, I'd be a little leery of walking out over the moors alone at night, wouldn't you?

CHAPTER THREE

The Sasquatch

He rises slowly, deliberately from his haunches and heaves himself fully erect. On his feet, he stands nearly ten-feet tall and weighs close to twelve hundred pounds. Forget about Bambi's dad—*this* guy is the real "monarch of the forest." No one and nothing can challenge him. Breath huffs hotly from large, widely-spaced nostrils, steaming and rising in the coolness of the morning air. He smells wood smoke and food, and he wants food, wants it *now*.

Long, shaggy, gray-brown hair covers him from head to foot. Small amber eyes peer suspiciously and near-sightedly from under an extremely prominent brow ridge. He is nearly always solitary unless a female presents herself, and then he will mate vigorously, repeatedly, and with great pleasure. Their relationship won't be a familial one; he takes no part in raising or caring for his young. He's always been more a love-'em-and-leave-'em kind of guy.

A solitary man, oblivious, fries a pair of fresh-caught trout in bacon fat, the snap and sizzle of it in the iron pan drowning out other noises. The man has come to trap in these woods—beaver pelts, mink, and fox bound for the wealthy, insatiable European market. The horse cropping grass nearby stamps its feet nervously and flashes the white rims around its wild eyes. There finally penetrates the man's less-acute senses a soft snuffling sound accompanied by an unpleasant, all-encompassing odor. Looking up from the black skillet, he turns to see a tall, hairy man standing some thirty feet away peering steadily at him. Jumping onto the horse's back in a single, fluid bound, the trapper rides as if his very life depends upon it, as indeed it may. For many seasons, around the fire on long winter nights, he will tell of this encounter with the hairy man of the forest. Some will even believe him.

The Sasquatch, known more familiarly to his many friends and admirers as Bigfoot, has undergone something of a transformation over the years since sightings of the reclusive, hairy humanoid of the Pacific Northwest and Canada were first widely published around 1920. The term Sasquatch is derived from the Halkomelem word sásq'ets. The Halkomelem people are part of the First Nations from British Columbia, Canada. The Lummi people of the Northwest tell stories of Ts'emekwes, a large, hairy man.

Tales very much like these have arisen wherever humans live on the planet. The hairy, wild man has different names in different cultures: Amomondgo in the Philippines, Barmanou in Asia, Batuotut in Borneo and Vietnam, Bukut Timah in Singapore, De Loys' Ape in Columbia, Am Fear Liath Mòr in Scotland, Mohoao in New Zealand, Maricoxi in South America, Orang Mawas in Malysia, Orang Pendek in Sumatra, Shōjō in Japan, Tsul 'Kalu in the American West, Urayuli in Alaska, Yeren in China, and Yowie in Australia are just a sampling of them.

By the late 1920s, reports taken from indigenous populations about the existence of a large, hairy man of the woods and gathered by John W. Burns were published in Canada's *Maclean's* magazine, now a weekly current-affairs periodical. Burns, an Indian agent and teacher, interviewed members of the Sts'Ailes people around Chehalis, Washington. They reported to him that the Sasquatch are able to speak Ucwalmicwts, the language spoken around Port Douglas, British Columbia, near Harrison Lake. Burns was the first writer to use the term Sasquatch in describing these beings. Before his stories were published, nearly all sightings occurred in the northwestern United States, mostly in Washington state and Western Canada. Afterwards, sightings began cropping up across the North American continent.

Human nature being what it is, hoaxes abound. It's easy is to suit up your buddy in an ape costume and film him lumbering across the road and into the woods. Bingo—instant payoff. Where there is money to be made, and it's most definitely very *serious* money, you will have hoaxes. Giant footprints are easily faked, too. Tourist dollars, royalties, and personal notoriety are all inducements for perpetuating the mythology of Bigfoot. Not all reports, however, have been made with an eye toward making a quick buck.

PR Newswire, on May 4, 2016, reports that the most famous hoax, the one of the creature walking through a clearing, was finally put to rest:

> Often-referred to as the holy grail of proof of Bigfoot's existence, the Patterson/Gimlin footage is now, for the first time ever, conclusively exposed as a fake in the new ground-breaking live radio broadcast: *Hoax of the Century.*

> Finally putting the nearly 50-year hoax to rest, renowned Bigfoot researcher Tom Biscardi presents testimony from those involved in staging the film, providing decisive evidence that the footage was not of some unexplained and mysterious natural being, but of a man in a suit. This landmark radio program will irrefutably prove that the Patterson/Gimlin film was perpetrated upon the public as a hoax for monetary benefit.

The wearer of the infamous suit, Bob Heironimus, was interviewed by reporter Richard Lei of the *Washington Post* in an article published on March 7, 2004. Heironimus "makes his 'full confession,' as he calls it, in a just-published book by paranormal investigator Greg Long, "The Making of Bigfoot." Long spent four years investigating the 60-second film clip and the people behind it. He traces the shaggy Bigfoot costume to a North Carolina gorilla suit specialist, Philip Morris, who says he sold it for $435 to an amateur documentary maker named Roger Patterson," for a hoax to be staged near Bluff Creek in northern California. Long told the *Post*: "Patterson was the cameraman. They made a gentleman's agreement that Bob would get in the suit and walk in front of the camera for $1,000." Heironimus' story is a bit different: "I was never paid a dime for that, no sir. Sure, I want to make some money. I feel that after 36 years I should get some of it. I only want what's coming to me."

Most hoaxes are created to garner fame or money. For Heironimus, the retired Pepsi bottler from Yakima, Washington, who wore the monkey suit, it comes down to the latter.

Many people who honestly believe they have seen the Sasquatch have, in fact, more likely seen one of the species of bears native to the region. Bears, particularly *Ursus arctos horribilus,* the "terrifying bear" identified by George Ord in 1815, seem particularly likely to be misidentified as a giant, hairy man. The grizzly bear is a subspecies of the more common brown bear. Grizzlies stand on hind legs, are as tall as six-and-a-half feet, and weigh up to 800 pounds. Males, however, have been found as tall as nine-and-a-half feet and weighing about 1,500 pounds, a considerably *big* Bigfoot, indeed! The bears rise up on their back legs and reach as high as possible to scratch their four-inch claws, claws as sharp as Bowie knives, high across the trunks of trees to warn competitors to keep out of their territory. Such bears have five toes and large claws, just like some of the plaster casts made of supposed Sasquatch tracks.

A prominent brow ridge often used to describe the face of the Sasquatch matches that of a grizzly, too. Just like the legendary hairy man, the grizzly usually prefers to avoid any contact with humans as much as possible. It generally only attacks when protecting its source of food or its cubs. Grizzlies came across the Bering land bridge along with humans into Alaska from Asia where the bears originated about 50,000 years ago.

Chief Etsowish-simmegee-itshin, or "Grizzly Bear Standing," was a chief of the Pend d'Oreilles in the first half of the nineteenth century. The group lived in British Columbia before Western Europeans began to move into their land. Seeing a grizzly bear on his hind legs is one of the most formidable and terrifying sights in the forest, one that would undoubtedly be imprinted deeply in the memory of any person who saw it. Small wonder that a strong leader would appropriate the powerful image. Stories of such sightings would naturally be passed down, embellished upon, and shared over the years. It is fairly easy to imagine these sightings evolving into the legend of the Sasquatch.

Some propose *Gigantopithicus*, a large, extinct primate, as a possible candidate for Sasquatch, although there is no evidence that this extinct Asian animal walked on two legs nor have fossils of it ever been found in North America. Scientists agree that it was a quadruped, not a biped. Still, its enormous dimensions are quite similar to the largest of the grizzlies. *Gigantopithicus*, if he could have risen upon hind legs, would have stood over nine-and-a-half feet tall and weighed perhaps 1,200 pounds, three or four times as much as a modern gorilla. Only teeth and mandibles have been found, some in Chinese apothecary shops where they are ground up and used as medicine. *Gigantopithicus* didn't become extinct until 100,000 years ago. It lived in Vietnam and Southeast Asia. Could a remnant population of *Gigantopithicus* have crossed the land bridge? It's definitely a possibility, but without any physical evidence to support it, this idea is simply speculation.

Extinct hominids such as the Neanderthal, *Paranthropus robustus*, and *Homo erectus* have been put forward as possible contenders, as well. Each would seem logical if their fossil remains had actually been discovered in North America, but they have not. That isn't to say the fossils aren't here waiting to be found, but again, without any evidence, the idea is implausible.

Jo-Jo the Dog-Faced Boy was a popular sideshow attraction displayed by P.T. Barnum, the father of humbug, from the time the boy was sixteen years old. His actual name was Fedor Jeftichew, a Russian born in 1868 who spoke many languages. He had a medical condition called hydrotrichosis that caused nearly every part of his entire body to be covered with long hair (not the soles of the feet and hands nor the mucous membranes). The genetic trait is passed through families. It may be present from birth or develop later. Fedor's father, Adrian, also exhibited the condition. Some families have several individuals who exhibit the condition and others without it but who still carry recessive copies of the gene. Because of the appearance of affected individuals, it has been called Werewolf Syndrome. The genetic mutation is extremely rare. Hydrotrichosis and a similar condition called hirsuitism (excessive hair growth, often in women with a surplus of testosterone) don't occur often, but when they do manifest, it's not difficult to imagine affected individuals during less enlightened historical periods being driven from their communities to eke out a living on the margins of society in order to avoid persecution. Because every continent except Antarctica has folk stories of these hairy men or wild men, it's not unreasonable to think that *something* must have prompted them.

Sasquatch societies and pseudo-scientists work diligently to perpetuate stories about Bigfoot, going so far as to stage "sightings" and take bogus "footprint" casts for profit. Those, we can discount. Unless or until the fossil record corroborates speculation about remnant populations of early primates or humanoids that first migrated to and then survived in the Pacific Northwest, those hypotheses must be discounted, too.

People of the First Nations of Canada and the Native Americans of the Pacific Northwest, however, had no such financial inducements. They didn't sell their stories to sensationalist periodicals or profit financially by their sightings of large, hairy "men" in the forest. For now, the majestic and mighty grizzly bear remains the most likely candidate for providing the inspiration for the stories of the Sasquatch. That isn't to say a genuine Sasquatch won't be discovered at some future point in time. I'd be the first one to stand up and applaud. Until then, however, I'm sticking with the grizzly hypothesis.

CHAPTER FOUR

The Chupacabra

It is alone and cold on this bleak night in mid-December, loping over the endless sands with no coat of hair to keep it warm. The primary need, the need to fill its belly, urges it forward. It not only wants to eat, it wants to *kill*. Normally, it would prefer to avoid anything and any place that smells of human—and this place reeks of men. Humans bring death with their guns, their poisons, their traps, and even their vehicles on the wide ribbons of asphalt that cross and recross the desolate landscape. It has little choice, now. Food is scarce and competition for it intense. Stronger, larger predators may take what they want, and that is why it finds itself skulking about in the shadows listening to the quiet sounds of prey seductively calling to it from just on the other side of the flimsy wire fence.

Uneasy clucks come from fat chickens who sense its presence and worry. Soft bleats from sleepy little goats and woolly young sheep promise warm, vulnerable bodies almost begging to be tasted. First and foremost, it is a predator. Long fangs and claws equip it to hunt the defenseless plant-eaters; the presence of so many, so close, incites it to boldness. It bunches the strong muscles of its rear haunches and propels itself over the fence easily. Immediately, the sounds of the animals it smells and hears become louder, more insistent, fearful. They start to panic. The fretful sounds they make only embolden it.

In a frenzy of killing, it takes and shakes one after the other in quick succession, sharp incisors and curved canine teeth piercing the soft flesh of their necks, reveling in the raw taste of life as it leaves their bodies. Blood flows wet and dark, staining its muzzle. Ecstasy. The final moments, just before a bullet pierces its skull, are absolutely blissful.

Stories of blood-sucking animals draw their inspiration from vampire mythology. Any animal driven by bloodlust but not easily or immediately identified is now called a Chupacabra. The name in Spanish means goat-sucker, *cabra* for "goat" and *chupa* for "it sucks." In a lot of ways it does—suck, that is. Ground zero for the Chupacabra, a very modern cryptid, can trace its origins to a report published in August 1995 in Puerto Rico. Eight sheep were killed in March of that year. Next came an attack killing as many as 150 animals near the same area in August.

A single woman, Madelyne Tolentino from Canóvanas, gave her account of the beast, and from there the story spread like wildfire. Also in 1995, on July 7, and *not* coincidentally, the film *Species* was released starring Academy Award-winning actor Ben Kingsley. A creature in the film called Sil perfectly matches Tolentino's description. It had long, powerful legs that allowed it to jump long distances. It stood between three and five feet tall, had red, glowing eyes, claws, an offensive odor, and a pronounced spinal ridge with spikes down its back. It was also hairless and dark grayish in color. In many ways, it resembled a bipedal reptile. The similarities between the single eyewitness account of the Chupacabra and the filmed, fictional monster were a perfect match. The creatures might have been identical twins. Tolentino admitted to seeing the film herself. She was apparently confused to some extent (if we're being charitable about it) and thought the filmed incidents were a sort of documentary.

Five years after it first made its appearance, the descriptions of the Chupacabra began to change. Instead of reptilian, it became a canine-type creature. It still retained some of the original characteristics, true, but the animal was undergoing a distinct metamorphosis. Red eyes and a foul odor remained, dark skin remained, but now it went on four legs instead of two. The creature preyed upon livestock, mostly in Spanish-speaking areas in the southwestern United States. Small puncture holes were found in the necks of goats and sheep. Chickens and other fowl were slain in large numbers. People's pets weren't safe. Something was very definitely attacking and killing domestic animals. Reports came in from Maine, Texas, Russia, Chile, and the Philippines. It was then that carcasses of the Chupacabra began to appear.

From there, it was a simple matter to determine the animal's DNA. This was a mystery easily solved, in contrast to other legendary monsters whose identities continue to remain elusive. The strange carcasses turned out to be those of dogs, coyotes, foxes, and even raccoons. Coywolves, hybrids of a coyote and the longer-legged wolf, and coydogs, a coyote-dog hybrid, also exist in the wild. These hybrids are often much larger than coyotes. The dead animals were gaunt to the point of emaciation, an unfortunate circumstance that made their legs seem unnaturally elongated. This also caused their eye sockets to protrude and their backbones to be visible with some tufts of hair resembling spikes running down the spine. The carcasses of the Chupacabra all had one thing in common: a severe parasitic mange infestation.

Demodectic mange, a parasitic infestation of the *Demodex* mite, may be localized, affecting only certain areas of the body, or it may be generalized, in which case it affects the afflicted animal's entire body. That's the reason why the animals perpetrating the attacks were hairless and smelled bad. They were very sick.

Sarcoptic mange, a condition caused by the *Sarcoptes scabiei* mite, is another skin disease found in dogs. These mites burrow through the skin and cause extreme itching and irritated skin. Repeated scratching that comes as a result of the mange causes the majority of the animal's hair to fall out. Skin rashes and crust formations on the affected areas can occur. It is extremely contagious and can rapidly spread through a population of canines living closely together in a pack.

Ker Than, writing for the *National Geographic News* in an article published on October 30, 2010, reported that a wildlife-disease specialist named Kevin Keel has seen images (and they are widely posted online) of an alleged Chupacabra corpse, He immediately recognized it as a coyote, but said he could imagine how others might not. "It still looks like a coyote, just a really sorry excuse for a coyote," said Keel, of the Southeastern Cooperative Wildlife Disease Study at the University of Georgia. A layperson, he noted, would probably not recognize the species. Keel goes on to say that University of Michigan entomologist Barry O'Connor has speculated that the mite passed from humans to domestic dogs, and then on to coyotes, foxes, and wolves in the wild. In humans, the infestation is called scabies, a manageable, non-life-threatening condition passed from person to person by close contact. The parasitic mites, of course, have nothing to gain by killing their human host.

Other species of animals, however, have not had time to adjust to the mites and suffer far worse when they contract the illness. O'Connor's research indicates the reason for the different responses is that humans as well as other primates have lived with the eight-legged, microscopic *Sarcoptes* mite for much of their evolutionary history, while other animals are only recently becoming exposed. "Animals with mange are often quite debilitated," O'Connor told reporter Ker Than. "And if they're having a hard time catching their normal prey, they might choose livestock, because it's easier."

The animal victims of the Chupacabra, supposed to have been completely drained of blood, were found to have plenty of blood left, upon examination by qualified professional veterinarians. The characteristic puncture wounds that killed them, mostly administered in their necks, were typical of throat wounds inflicted by dogs and coyotes who instinctively go for the jugular vein in their attacks. Neither is it at all unusual for such animal attacks to leave victims whole and uneaten. Coyotes and dogs sometimes kill repeatedly for no apparent reason other than the sheer sport of it.

Even if mange is not the culprit, there exists an unusual variety of dog native to northern Mexico called the Xoloitzcuintli, or Xolo for short. It is also called the Mexican hairless dog. The breed comes in toy, miniature, and standard sizes, all three of which can be born within the same litter. Some are born with short, dense coats, and others are virtually hairless. It looks very much like the ancient Egyptian pharaoh hound. The standard Xolo is dark gray or black in color, has large, bat-like ears, and a whip tail. The shape of its head and ears conforms with some of the descriptions likening the Chupacabra to a devil bat. While it is the national dog of Mexico with roots going back 3,000 years to the Aztecs, Toltecs, and Mayans, specimens of this relatively rare breed are not generally recognized by most people in the United States, a fact which might lead to it being confused with a Chupacabra.

Similar stories predate the Chupacabra by many centuries. The bat god Camazotz was feared by the Zapotec people of Oaxaca in Mexico. The K'iche' group of Mayans in Guatemala told tales of dark, bat-like reptiles that heralded death. As long as there have been people, there have been tales of "long-leggedy beasties, and things that go bump in the night." If the modern legend of the Chupacabra is any indication, there always will be.

CHAPTER FIVE

The Loch Ness Monster

The solitary hiker sat on a tufted hillock overlooking the deep blue of Scotland's scenic Loch Ness. A little spotted spaniel flopped companionably down beside him. He drank from a canteen of water and gave some to his dog in a shallow dish brought along for that purpose. They shared a modest meal of bread and cheese. Tonight, maybe he'd manage to hook a trout for supper. They'd been traipsing about the countryside for several days.

The young man, barely out of his teens, was on holiday from his position as a bookkeeper in Glasgow and found the loch to be one of the prettiest sights he'd come upon. This was a wild country with its craggy tors and desolate, wasted moors, its chill dampness that cut to the bone and shrill winds that mercilessly buffeted the terrain. Nevertheless, it was a welcome change from the cramped, smoky old city. The heather wouldn't bloom for a few weeks yet, as it was late May; everything around remained gray except for the deep green of the forests and the deep blue of Loch Ness. His ramble had taken him to Inverfarigaig, a pretty, picturesque spot where he planned to camp for the night. In the morning, he'd venture beyond to the Falls of Foyers.

Yap-yap-yap-yap came the frenzied, high-pitched barking.

The dog sprang up with four feet planted. The hackles along her back and neck bristled stiffly. The man looked around and saw nothing that might have elicited such a response from his good-tempered dog, neither squirrel, waterfowl, stray sheep, nor some other hiker.

A splash drew his attention back to the loch where something large and dark had broken the surface and was cutting rapidly through the water. If it *was* a salmon, it was an incredibly big one. He scrambled for his fishing pole to make a cast in hopes of catching the monster fish before it swam away.

A gracefully curved neck, almost like a swan's, rose in front of the quickly moving hump. From the rear, a long tail thrashed about. The head, dripping water, was smallish in comparison to the body, and serpent-like. Firmly clasped between its sharp teeth, the man could see pinioned the big salmon he had hoped to catch. With a practiced toss, the creature flung the large fish upward and took in down in one, massive gulp. It propelled itself forward with four flipper-like appendages that reminded him of those of a dolphin, except the flippers on this beast were many times bigger than that. Within the span of a few seconds, it sank back into the loch, dark water closing over it soundlessly.

The spaniel continued to bark furiously for several minutes. "Hush, Gracie, hush, little girl." He soothed the spaniel with a few gentle pats. Neither one of them could tear their eyes away from the spot where the odd creature disappeared. "I saw it," he told her, for there was no one else to tell—and who would have believed him, anyway?

The Loch Ness Monster has a long and illustrious history. The lake has been the source of study for decades, and sightings of the water-beast have been reported over the past 1,500 years, practically ever since there have been people living near the lake. Our modern preoccupation with Nessie, as the monster is familiarly known, began on May 2, 1933, when the *Inverness Courier* printed the story of a local couple who claimed to have seen "an enormous animal rolling and plunging on the surface" of Loch Ness. Immediately, the "monster" became a media sensation. London newspapers sent reporters to Scotland and put up a 20,000 pound sterling reward (about $1,657,000 in today's dollars) for capture of the animal. At nearly 800 feet deep and 23 miles long, Loch Ness is the largest body of fresh water in Great Britain. Is this sustained hoopla *all* hype? Until something verifiable is either caught or convincingly caught on film, there's no way to know for sure, but some theories are more intriguing than others.

Tribes of Picts living in the area around 500 BC carved a strange-looking aquatic creature into standing stones near the loch. The first written mention of a lake-dwelling monster appeared in a seventh-century biography of Saint Columba, the Irish missionary who introduced Christianity to Scotland. In 565 AD, his biographer wrote that while Columba was traveling to visit the king of the northern Picts near Inverness, a series of murderous aquatic animal attacks occurred around Loch Ness. The writer claimed that St. Columba invoked God's name, challenged the killer beast to end its murderous spree, and ordered it go back into the lake, and stay there. According to his biographer, it did.

London's *Daily Mail* hired a big-game hunter, Marmaduke Wetherell, to find the elusive Nessie. He discovered footprints, while another local couple said they had seen the creature lumbering across the road. The newspaper claimed "Loch Ness Monster is not Legend but a Fact," that is, until the footprints were determined to have been created using a dried hippopotamus foot, the kind routinely used for umbrella stands in Victorian times. Interest in Nessie subsequently waned.

The iconic 1934 photo of Nessie, widely known as "the surgeon's photo," showed a head on a long neck rising out of the waters of the loch. It created a new frenzy of international attention. The animal certainly *looked* like a prehistoric creature. It was a deliberate duping of the public perpetrated by Marmaduke Wetherell to take revenge for his stinging humiliation caused by the earlier hippo tracks hoax. His adult son later recalled Wetherell saying, "We'll give them their monster." You can almost envision him twirling his mustache as he said it and emitting an evil chuckle. British surgeon Colonel Robert Wilson was the "front man," a respected professional who lent the scam credibility. Christian Spurling was the artist who made the small model of a clay figure mounted on a toy submarine. At the age of 90, shortly before he died, Spurling confessed that he made the model. The photo was cropped to make the monster appear larger, but by 1984 the hoax had already been exposed by Stewart Campbell in the *British Journal of Photography*. The object could not have been larger than two- or three-feet long, Campbell concluded, based on the wave patterns in the water. That 1934 photo, though, was viewed by many, my highly intelligent father included, as definitive proof of the Loch Ness Monster's existence.

Science tried to end the controversy. In 1975, Boston's Academy of Applied Science made an expedition to the loch using sonar and photography. One photo appears to show a giant flipper like one you might see on a living plesiosaur or an elasmosaur. It appears quite convincing. This photo was taken nine years before the surgeon's photo was exposed as a deliberate fraud.

Other expeditions in the 1980s and 1990s used sonar. Those findings were inconclusive. The readings indicated something was submerged, but there was no proof of what that might be. Regardless, people continue to haunt the banks of Loch Ness. They continue to scan the expanse of the big lake hoping against hope that a monster from the distant past might just surface. If it does, they'll be ready with their cell phone cameras to record the event.

There was a good deal of excitement generated when a film prop, located by sonar, was found at the deep bottom of the lake. A robot called Munin was used by a survey team from the Norwegian company Kongsberg Maritime to monitor the lake bed. This joint project was undertaken by the tourism organization VisitScotland, a fact that taints the entire enterprise with bias. The beast had two humps and looked briefly promising as a monster candidate. It turned out to have been something created for *The Private Life of Sherlock Holmes* (1970) directed by Billy Wilder. It was sunk in the lake after the filming wrapped. Again, widespread disappointment and a gusty, collective exhalation of breath from Nessie's legions of ardent fans followed the announcement.

As I write about Nessie's history, a woman visiting the loch with her family has just taken a photo of a fin rising from the water. It looks like a dolphin, says Martin Little, writing for the *Press and Journal,* on August 30, 2016. Here is his hilarious, hedging headline: "Is it the Loch Ness Monster? Or is it a dolphin? (… probably a dolphin)." There is speculation that fishermen may have netted a saltwater dolphin with their catch and released it into the freshwater lake as a prank. The newspaper described it as a "clearly defined fin with a white tip [that] can be seen breaking the surface of the water."

Tony Drummond has been working on Loch Ness for ten years. A few days after the dolphin photo was taken, he saw a "white-streaked" creature swimming under the water, but Drummond doesn't believe it was a dolphin. He thinks it might have been a large salmon. He told reporter Martin Little: "It did not break the surface and it was hard to tell its size, but it was about 3.5 feet by two feet. A sort of oval, coffee-table size. It was moving, but not fast. It appeared about 15 metres away to the right. It was under the water with a bit of white in it. I have never seen anything like it in my life. But I will eat my woolly hat if it's a dolphin. It could be a large salmon—I've seen them around 100 pounds in here."

Another speculation about what Nessie could be is duller than a dolphin. Victorians released a species of very large catfish into the loch. "The Wels catfish, also called sheatfish, is native to wide areas of central, southern, and eastern Europe, and near the Baltic and Caspian Seas. The giant, fearsome fish can grow as long as 13 feet and up to 62 stone (886 pounds)," writes Little. Other sources estimate a maximum body length of 16 feet. It is exceptionally bizarre-looking. Wels catfish can live up to fifty years, and there have been a few reported attacks by these giants on humans. To be fair, in those cases it was the person who went after the catfish first, and not the other way around. If you came face to face with one of the biggest specimens, you could certainly be convinced you'd seen a monster!

Steve Feltham has been recognized by the *Guinness Book of Records* for maintaining the longest sustained hunt for the Loch Ness Monster. He has been a regular fixture in the area for decades and is an authority on the topic. He has come to the conclusion that Nessie is quite likely a Wels catfish.

It is definitely in Scotland's financial interest to keep the legend of the loch alive. Nessie is estimated to add £25 million (almost $33 million) in tourism to the local economy every year. There are certainly large, aquatic creatures swimming in Loch Ness. For my father's sake, I hope one day Nessie will rear her pretty head and prove he was right after all.

CHAPTER SIX

The Werewolf

It was a crisp fall evening, the kind that made you feel glad to be alive. The golden orb of a harvest moon peeped over the dark, reaching-finger silhouettes of the treetops. She inhaled deeply, enjoying the sharp smell of burning leaves in the air. In her hand she swung a hamper that contained some home-baked treats and a bottle of good Napa wine. She knew her ailing grandmother would love having her stop by to visit, as much as she would enjoy the goodies in the basket. The elderly lady was recovering from a bout of pneumonia and hadn't been able to get out of the house for several days. A stiff wind picked up as the temperature dropped. Good thing she'd thrown on her favorite crimson wool jacket and cashmere scarf, a birthday present from Gran. She was thinking no further ahead than the visit, choosing her path carefully as the light faded.

The scent on the wind was delicious. No, not the overpowering, sickening smell of the fudge brownies and blueberry muffins. That was merely annoying, just like the flies that would come to feast on the carcass left behind once he had eaten his fill. The sugary stench competed, along with some kind of disgusting floral perfume, with the tender sweetness of the young girl's ripe, rosy flesh. He hadn't eaten in weeks. Saliva already overflowed his lips and hung down in foamy strings from his jaw. It was an involuntary response to the nearness of the succulent prey and his sharp hunger pangs. Every muscle tensed as he prepared to attack. Between his shoulders, they bunched into knots. His nose quivered. His feet danced up and down in almost-giddy anticipation.

She heard some slight rustling in the fallen carpet of leaves just to the right and slightly ahead of her in the woods. Probably a squirrel gathering a few more acorns to store for winter. Maybe a stray cat or dog had wandered too far from home. She felt as safe as houses this evening. It was a path she'd followed over and over since she was very small, and it was as familiar to her as was the way to her own home. The persistent, niggling noise continued. Perhaps an owl had caught a small, helpless creature that had ventured too far from its burrow. Whatever it was, the noise was still there. She frowned and pensively bit her lower lip.

Giving no warning growl of intent, the creature closed the short distance between them in two mighty bounds. The hamper thudded to the ground, spilling the contents across the path. Lesser animals would dispose of those easily enough. The vinegary stink of fermented grapes was soon absorbed into the dirt and dry leaves when the bottle that held it shattered against a rock. Before the girl could even raise her arms to defend her vulnerable throat, he had ripped it away with one snap of impossibly strong jaws lined with sharp, ivory-colored fangs. Blood gushed warm and red over him and her, joining them as one in the ancient dance of ecstasy—well, ecstasy for *him,* at any rate.

Transformation legends are as old as recorded history. Human beings somehow shape-shifted into other animals. Warriors would draw the strength and power from a wolf if they wore its skin. The bite or scratch from a werewolf would turn the injured person into one himself. In modern times, the werewolf has been domesticated. No longer the personification of unmitigated evil, a werewolf is depicted as strong, seductive, and even sexy. Charlane Harris' *True Blood* series features a very attractive werewolf who also happens to tend bar. *Teen Wolf, Bitten, Lost Girl,* and *Buffy,* TV dramas, also showcase handsome or beautiful werewolves living uneasily, although very attractively, among humans. Jack Nicholson in the 1994 film *Wolf* gives a particularly memorable performance as the title character. Stephen Sondheim's *Into the Woods* upends the familiar Little Red Riding Hood story into one of overt sexual awakening as the Wolf smoothly croons "Hello, Little Girl". How did we move from fear and loathing to fascination and attraction? The path was a long and bumpy one.

From German paganism, with its origins in early European legends, the werewolf mythology spread to France and then to England. Slavic and Baltic cultures have their own traditions, independent of those based in Germany and Western Europe.

The werewolf is referenced in classic literature from Greece and Rome. Lycaon served human flesh to Zeus in one story, and was turned into a wolf for it. From his name we get the word lycanthrope, meaning werewolf. The ancient Germanic legends held on longest among the Vikings who believed that men who wrapped themselves in the skins of wolves would be able to harness the power of a ferocious wolf in battle. From Scandinavia, the stories are thought to have spread into Russia. Asian cultures had similar tales, but in the east, humans were thought to transform into leopards or tigers, not wolves.

In an odd side note, people believed it was possible to detect a werewolf while in its human form based on certain physical characteristics, the unfortunate possession of the unibrow being one of them. Other tip-offs were low-set ears and a peculiar rolling gait. In some regions it was believed that if a werewolf in human form were cut, telltale fur would be present in the wound.

In the Late Middle Ages, Switzerland became the epicenter of witchcraft trials. They began in 1428 in Valais before finally sputtering to an end in the 1750s. There was a good deal of conflict between orthodox Catholics and Protestants with opposing viewpoints, and also between Christianity and Paganism, some of which naturally spilled over into the trials. Along with accused witches, werewolves were also tried, although never in such great numbers. Scholars of the subject estimate that between 40,000 to 60,000 witches were executed. Records show just 280 trials of werewolves were held. From Switzerland, fear of the occult and Christian-led trials against those who would practice it spread across Europe. Colonists carried their beliefs in witches and werewolves along with them to the New World. There are two main reasons for the existence and persistence for a belief in the existence of werewolves.

The first of these is the aberrant, psychopathological behavior of certain individuals afflicted with strange and terrible compulsions. This type is perhaps best illustrated by the case of a wealthy Rhemish farmer, Peter Stumpf, although there are various other spellings of his name. The evidence given in his 1589 trial is critical to any understanding of the werewolf phenomena. He was a widower with a teenaged daughter and a young son. He had taken a distant relative as his mistress, breaking a long-standing cultural taboo. Although no original papers of the trial remain and the German accounts have been lost, there are two copies of English pamphlets detailing the gristly events. If this case isn't enough to give you nightmares, I don't know what would.

While being horrifically tortured on a wheel, Stumpf confessed that from the age of twelve, he practiced black magic and made a compact with the Devil. When he put on a belt the Devil gave him made of wolfskin, he had the power to transform into a wolf himself. The old pamphlet says that he took on "the likeness of a greedy, devouring wolf, strong and mighty, with eyes great and large, which in the night sparkled like fire, a mouth great and wide, with most sharp and cruel teeth, a huge body, and mighty paws." As soon as he took off the belt, he would again resume his human shape.

Peter confessed to eating raw flesh of animals, and when that was not enough satisfy his blood-lust, he consumed fourteen children and two pregnant women, along with their unborn babies, whose still-pulsing hearts. he said, were "tasty morsels." He also ate the brains of his own son, one of the fourteen victims, and slept with his teenaged daughter and the female relative. Both of those crimes were considered incestuous. Following the trial, the two women were killed along with him as being "complicit," although they were his victims.

The execution took place on October 31, 1589. It was just about as grisly as his crimes and was obsessively thorough. He was tied to a wheel and pieces of his flesh were removed with red-hot pincers in ten different places. Next, his limbs were removed and crushed to prevent him from coming back from the grave and resuming his notorious killing spree. His body, or what still remained of it, was burned along with the bodies of his unfortunate daughter and the female relative. The women had been flayed and strangled first. Finally, his head was cut off and mounted on the top of a pole. The wheel upon which he was tortured was also tied to the pole, along with the figure of a wolf. It was put on display to dissuade others who might seek to follow Peter Stumpf's awful example. Jeffrey Dahmer, among others, clearly didn't get the message.

A second explanation for the widespread and geographically diverse legends of the werewolf is that the person suffers from an actual medical condition. Hypertrichosis, the growth of hair upon most surfaces of the body, has already been described in the previous Sasquatch chapter of this book. Look at paintings and photos of those who exhibit the condition and you will be immediately struck by their close resemblance to infamous werewolf Lon Chaney, Jr., in the iconic horror movie, *The Wolf Man* (1941). The budget was $180,000, the special effects rudimentary, but the metamorphosis from man to wolf results in a likeness so similar to individuals with hypertricosis that it is hard to imagine that affliction wasn't the make-up artist's inspiration. Petrus Gonsalvus, in a portrait called *The Hairy Man* painted by Joris Hoefnagel (1542–1601) from his "Elementa Depicta," looks as if Chaney himself might have posed for it.

Another intriguing theory, put forth by Lee Illis of Guy's Hospital in London, suggests a medical condition called congenital porphyria as the inspiration for the werewolf legend. The incidence is rare, and is triggered by a recessive gene. In 1963, Illis wrote a paper titled "On Porphyria and the Aetiology of Werewolves" that was published in *Proceedings of the Royal Society of Medicine*. "A belief as widespread in both time and place as that of the werewolf must have some basis in fact. Either werewolves exist or some phenomena must exist or have existed on which, by the play of fear, superstition, and chance, a legend was built and grew." People with porphyria have photosensitivity and are unable to tolerate sunlight. They may also suffer from hypertrichosis, excessively hairy skin, and skin pigmentation. Untreated, the illness results in reddish teeth which might look as if they were covered in blood, and manifests itself in psychosis. Over time, severe lesions cover the body, disfiguring it. It wouldn't make a person look like a wolf, however. Similarly, porphyria has been used to explain vampire legends.

Ian Woodward disagrees, in his book *The Werewolf Delusion* (Paddington Press, 1979). He notes that werewolves looked like actual wolves, not people—not even very hairy people. He focuses on the fact that people bitten or sometimes even scratched by a werewolf would become a werewolf. For these reasons, he suggests the transmission of rabies from a bite as a source of the legends, but this twist is a relatively modern one. Many have agreed that rabies is a possible explanation, but the earliest known stories of werewolves do not include the cause of it as being bitten by an afflicted animal. If it did, we might have the legend of the weredog, the werebat, the wereraccoon, and the wereskunk, since many animals can contract the disease of rabies. In fact, England's Nick Park created a claymation feature called *Wallace and Gromit: The Curse of the Were-rabbit* (2005). It has been pointed out that being bitten by a rabid bat, however, might well be the source of vampire legends.

Yet another psychological condition may result in people honestly believing they have transformed into animals. The clinical name for it is body dysmorphic disorder, a term originated by Enrico Morselli in 1891. He called it a "form of insanity" and wrote that "when one of these ideas (e.g., the feeling of physical changes that exist only in the person's imagination) occupies someone's attention repeatedly on the same day, and aggressively and persistently returns to monopolise his attention, refusing to remit by any conscious effort; and when in particular the emotion accompanying it becomes one of fear, distress, anxiety, and anguish, compelling the individual to modify his behaviour and to act in a pre-determined and fixed way, then the psychological phenomena has gone beyond the bounds of normal, and may validly be considered to have entered the realm of psychopathology."

The most notorious case of BDD was a Russian aristocrat referred to in the literature as the "Wolf Man." He dreamed of wolves in the branches of a tree outside his bedroom looking at him. He believed completely in purely imagined defects in his nose. He was treated by Sigmund Freud who, not at all surprisingly, decided that his nose represented his penis and that he wished it cut off so that he could transform into a woman—which seems like a pretty big leap to make from dreaming about wolves in trees.

Dr. Andrew J. Larger, a consultant neurologist at the Walton Centre for Neurology and Neurosurgery in Liverpool who maintains a particular interest in dementia and cognitive disorders, wrote in *Neurological Signs* in 2010, "I had always understood lycanthropy to mean the transformation of a human into a wolf (Greek: lukos-wolf, anthropos-man). Such animal-like behaviour has a long history. The mythical werewolves so beloved of the film industry aside, these cases, sometimes labelled 'clinical lycanthropy' to emphasize the distinction, usually seem to be associated with psychiatric disorders such as psychosis or depression and have been understood as delusional disorders in the sense of self-identity disorder." He describes the plight of Odysseus' men changed by Circe into pigs but notes that on the island were "mountain wolves and lions that she had bewitched with her magic drugs." People report being changed into a wide variety of animals, not just wolves. The transformation only happens in the minds and actions of the person with the illness. Clinical lycanthropy is thought to be a result of schizophrenia, clinical depression, or bipolar disease.

Mental and physical illness are the scientific explanations for the phenomena of the werewolf. Whatever the inspiration for the legends around the world, they have been with us since the beginning of recorded history, and, if anything, they are even more deeply entrenched in popular consciousness in modern times than they were in the past. Werewolves may not be actual creatures of the night, but as long as people spin stories about them, they are likely to remain a fascinating topic of speculation for years to come.

CHAPTER SEVEN

The Yeti

There weren't so many yaks in his father's herd that the boy didn't recognize and feel affection toward them as individuals. They sustained the family though those bitter winter months when the wind howled like a savage, living creature racing over the peaks and down the valleys of the mountains at the top of the world. They gave milk, meat, and finally their coarse hides to the people who tended them carefully. Their manure was used to fertilize crops and they were harnessed to work as draft animals. Without them, the people of the region would likely not have been able to survive.

Domesticated yaks are the descendants of wild yaks caught and tamed by late Stone Age people some ten thousand years ago. Yaks cannot survive below 10,500 feet above sea level. They thrive at 12,500 and have been found as high as 18,300 feet. The boy knew none of this. He didn't go to school, never had and never would. The winter camp was three days' walk from the nearest road, the summer camp farther still. He tended the herd lovingly and thought of the yaks as valued members of his extended family.

Snow blew horizontally this morning. The sun was up but almost totally obscured by a low ceiling of iron-gray clouds. The people knew how to survive here, as their fathers and grandfathers had from time immemorial. The yaks were tolerant, patient creatures. They were nearly covered in snow and looked like nothing so much as a grouping of small, white hillocks against the mountainside.

The treeline in the Himalayas is the highest in the world, extending to about 16,000 feet. From the edge of the trees, maybe twenty yards away, a tall man silently emerged. The creature stood nearly twice the boy's height and weighed many times more than the biggest man in the village. He was covered in long, shaggy hair and looked very fierce with long claws and a mouthful of sharp, pointed teeth. At once, the boy knew he was in the presence of the huge man of the forest. He froze, not only in fright although he felt plenty of that, but also in reverence. Most people in the village would live their entire lives without ever seeing this awesome sight.

The creature selected and then grabbed the largest of the spring calves with one of its strong arms and easily snapped its neck with a single swipe of a powerful paw. The poor animal's plaintive cry was immediately cut short. Others in the herd shuffled uneasily away, seeking some kind of reassurance from the human boy who was always with them. He called to the herd softly. The man of the forest had chosen his sacrifice. Seemingly without effort, the fearsome monster dragged the calf back with him into the woods from whence he had come and disappeared behind the curtain of snow.

Some sort of living creatures in the Himalayan Mountains have caused human beings for hundreds, even thousands, of years to believe in the existence of a hairy hominid dwelling in the snows at the highest part of the world. The Yeti is the cold-climate cousin of the North American Sasquatch. At times, it is benign and even helpful to people. At other times, it is menacing and dangerous. Because of the inaccessibility of its habitat, proving or disproving its existence has been extremely problematic. Until the middle of the nineteenth century, when New Zealander Sir Edmund Hillary and Tenzing Norgay, the Nepalese Sherpa who accompanied him, scaled Mt. Everest in 1953, the Yeti was simply a local story known to the very few individuals who inhabited the inhospitable region of the world's highest landforms in Nepal, Bhutan, and Tibet.

Even Adolf Hitler got in on the search for the Yeti, hoping to find the "missing link" between apes and men to bolster his theories on Aryan supremacy. Hitler financed Ernst Schäfer's expedition to Tibet. It was said to be for "zoological research" and was sanctioned by Heinrich Himmler, head of the SS and the Ahnenerbe ("inheritance from the ancestors"). The Ahnenerbe was a society created by Heinrich Himmler to scientifically prove Nazi pseudo-history. The hugely popular Indiana Jones films reference the Nazis' search for powerful religious artifacts. Himmler gathered some of the best experts in science, history, and anthropology in Germany to search for the Yeti. No such luck, but they *did* manage to steal some priceless Tibetan artifacts in order to avoid leaving the region entirely empty-handed.

There are many names for the creature. It is called Yeti in Nepal, loosely translated to mean "rocky place" and "bear." Michê is the Tibetan world for "man-bear." Bun Manchi is a Nepalese word for "jungle man." Mirka means "wild man," and Kang Admi is "snow man." Not until 1921 was it called an abominable snowman, a clever term coined by Henry Newman. That was the same year that Lieutenant-Colonel Charles Howard-Bury led the British Mount Everest reconnaissance expedition. In his book detailing the experience, Howard-Bury writes about crossing the Lhakpa La mountain at 21,000 feet. The party came upon footprints in the snow that he believed "were probably caused by a large 'loping' grey wolf, which in the soft snow formed double tracks rather like a those of a bare-footed man." Local Sherpa guides who accompanied him told him that they must be tracks made by a wild man of the snows, translated from their languages as a combination of "man-bear" and "snowman."

Henry Newman was a frequent writer for Calcutta's *The States-man*. He interviewed the porters of the British Mt. Everest reconnaissance expedition once they had returned to Darjeeling. He incorrectly translated their word "metoh," believing it was a synonym for filthy. The Filthy Snowman didn't pack quite the same punch as did the Abominable Snowman, so Newman took a bit of poetic license and substituted the world abominable and was the first person on record to do so. The appellation stuck, at least in the Western world.

Historically, people in the region before they become Buddists worshipped a god of the hunt they referred to as a glacier being. Others in the area called upon the Wild Man, a large hominid, in their sacred rites. In a chapter titled "Wild Men of the Mountains" in Laurence Waddell's book *Among the Himalayas* (1899), he observed that "some large footprints led across our track and away up to the peaks. These tracks were said to be those of the hairy wild men who are believed to live amongst the eternal snows...". Waddell discounts the reports. "The belief in these creatures is universal among Tibetans," he notes, but observes that the information is always second-hand, something repeated to the person he spoke with by another witness. The author concludes that the real wild men of the mountains "are actually the great yellow bear, *Ursus isabellinsus,* who are known to be carnivorous and often kill yaks."

With an increased Western presence on the mountains during the twentieth century, reported sightings of the Yeti or its footprints rose accordingly. The most famous footprint photos, those taken by Eric Shipton in 1951, were recently (in 2014) auctioned by Christie's and sold for five thousand pounds. From an article in the *Times of London*, Shipton said, "The tracks were mostly distorted by melting into oval impressions, slightly longer and a good deal broader than those made by our mountain boots. But here and there, where the snow covering the ice was thin, we came upon preserved impressions of the creature's foot. It showed three 'toes' (actually four) and a broad 'thumb' to the side. What was particularly interesting was that where the tracks crossed a crevasse one could see quite clearly where the creature had jumped and used its toes to secure purchase on the other side. We followed the tracks for more than a mile down the glacier before we got on to moraine-covered ice."

The tracks were estimated to be quite fresh, no more than a day old. According to Shipton, "Sen Tensing, who had no doubt whatever that the creatures (for there had been at least two) that had made the tracks were 'Yetis' or wild men, told me that two years before, he and a number of other Sherpas had seen one of them at a distance of about 25 yards at Thyangboche. He described it as half man and half beast, standing about five feet six inches, with a tall, pointed head, its body covered with reddish brown hair, but with a hairless face. When we reached Katmandu at the end of November, I had him cross-examined in Nepali (I conversed with him in Hindustani). He left no doubt as to his sincerity. Whatever it was that he had seen, he was convinced that it was neither a bear nor a monkey, with both of which animals he was, of course, very familiar." This account, unlike those shared with Waddell, was from a personal observation.

In addition to footprints in the snow, a Yeti scalp was found and tested. It was taken from the Pangboche monastery to be examined in 1954. The conclusion: it was neither ape nor bear but taken from the shoulder of a hoofed animal. In 1960, Sir Edmund

Hillary, who claimed to have seen mysterious footprints while scaling Everest, sent a Yeti scalp that had been on display in the Khumjung monastery off for testing. That scalp was determined to have been made from a serow, a Himalayan antelope. Even monasteries were not immune to deliberate deception.

American millionaire Tom Slick financed and led several expeditions to Nepal in the late 1950s looking for evidence of the Yeti. *U.S. New and World Report* published an article by Paul Bedard and Lauren Fox (September, 2011) that indicates how very seriously stories of the Yeti were taken by the American government in 1959. "'There are, at present, three regulations applicable only to expeditions searching for the Yeti in Nepal. These regulations are to be observed,' said a memo from the embassy written on State Department letterhead. The first rule required that expeditions buy a permit. The second demanded that the beast be photographed or taken alive. 'It must not be killed or shot at except in an emergency arising out of self defense,' wrote Embassy Counselor Ernest Fisk on November 30, 1959. And third, any news proving the existence of the Abominable Snowman must be cleared through the Nepalese government which probably wanted to take credit for the discovery." Slick's adventures are believed to have (very loosely) inspired the characters of Yukon Cornelius and the Bumble in the 1964 stop-motion animated feature *Rudolph, the Red-Nosed Reindeer.*

In 1983, Daniel Taylor and Robert Fleming, Jr. made serious progress in identifying the real basis behind Yeti sightings over the centuries. They focused their efforts on Nepal's Barun Valley. Footprints believed to be from the Yeti were discovered there in 1972. What Taylor and Fleming uncovered was enlightening. They found footprints which showed an animal who walked on two legs and made simple nests in the trees. Locals told the explorers of two kinds of bears living in the area. The rukh balu was smaller, about 150 pounds, and shy. It stayed mainly in the trees. The bhui balu was a different matter entirely. It was estimated to weigh 400 pounds and was aggressive, even vicious. Skulls of the two kinds of bears were collected and sent to the British Museum, the Smithsonian, and the American Museum of Natural History. But, there was just one bear. Both skulls were from the Asiatic black bear, *Ursus thibetansis*. The "two kinds" were merely those of young bears as opposed to mature ones.

The species is omnivorous. In form, it is very similar to a number of other prehistoric bears. According to the Denver Zoo, this species is the "most bipedal" of bears and has been "known to walk upright for a quarter of a mile." The zoo also notes that they "sometimes kill domestic livestock."

This would explain the dual nature of the Yeti, reported to be gentle at some times while savagely aggressive at others. It also accounts for the use of the word "bear" in many of the indigenous people's names for the monster. Several years of extensive research in the valley by Fleming and Taylor, aided by John Craighead (identical twin to fellow researcher Frank Craighead; the brothers are best known for their groundbreaking studies of grizzlies in Yellowstone) and Tirtha Shrestha, discovered that the Asiatic black bears take to the trees to survive when they are about two years old. Older male bears are known to attack and kill younger ones. While they live in the trees, the young bears manage to turn their inside claw away from the rest to act like an opposable thumb, thus giving them greater control as they climb. A footprint in the snow would show this outward-turned claw looking like a hallux, a big toe.

When viewing the famous 1951 Shipten footprint photos, it becomes pretty clear what it is they reveal. The prints provide evidence of the hind paw stepping directly into the print of the front paw as the young bear walks forward on all fours. It looks like a very long human footprint. With the footprints stepping back-foot-into-front-foot, the animal's gait appears to have been made by a bipedal, human-like creature. In fact, the tracks were quite probably made by the rare bears found in the area. Bill Garrett, then-editor of *National Geographic Magazine,* believes the bear-tracks hypothesis "gives us a believable yeti." So do I.

Tim Sullivan, writing for *The Victoria Advocate* in 2008, put it very well: "The high Himalayas are among the most isolated, forbidden parts of the world. Couldn't something—a species of gorilla or proto-human—have hidden for centuries amid the crags?" Sangay Wangchuck, Bhutan's director of Conservation with an MA from Yale and a Ph.D from the Swiss Institute of Technology, told Sullivan that without scientific proof of the Yeti, he cannot credit its existence. "Let's talk about it," he says, "but let's leave it at that."

CHAPTER EIGHT

The Sea Serpent

The young couple gazed deep into one another's eyes, and the world around them faded into a rosy haze. The setting was romantic, the most beautiful imaginable. The Golden Gate and the city skyline beyond it framed one of the most well-known sites in California or, indeed, the world. Lights began to switch on as day turned to night. It was sunset over the Pacific, they had picnicked on a grassy bluff high above the San Francisco Bay, and now they kissed. Tenderly, he traced the outline of her face as if he wanted to memorize it.

"I love you," he told her. It was the first time he had said it aloud.

"Me, too," she replied. Then, flustered, she explained, "I don't mean I love *myself*, I mean *I* love *you*, also. Um, I love you as well." She felt nervous, awkward, and silly, and she was talking too much.

The corners of his eyes crinkled as he smiled and then silenced her with another kiss.

Below them in the wine-dark waters, a slippery form rose from the depths and broke the surface. It wasn't a seal. It wasn't a shark, either, although great whites are occasional predators in these waters. Its sinuous head atop a long, slim neck was small, compact. It possessed a mouthful of needle-sharp teeth. A few loops of its elongated, supple body were now visible behind the head. Swiftly, it cut through the choppy waves at a good clip leaving the V of a foamy wake behind as it swam. Whatever it was, it was hunting.

The couple never even noticed.

As long as men have ventured forth upon the sea, they have returned to tell stories of the great and terrible serpents that lurk under the waves. Seafarers from Scandinavia seem to have been exceptionally favored with glimpses of numerous monsters over the centuries. Sighting one of the monsters was thought to be a portent—something big was about to happen, and that particular something wasn't usually good. They've appeared in the form of long, eel-like serpents with multiple humps. Some have been described as having a head shaped like a horse with a mane. Sometimes they swim like a mammal would, sometimes they move through the water like a snake. Still other times, their movement is compared to the odd, up-and-down locomotion of a caterpillar or inchworm. Witnesses characterize some of the animals as "playful," while others are said to be "menacing" and "dangerous." What are these creatures that appear again and again in the bays and open oceans of the world? Are sea serpents merely victims of mistaken identity, or does the possibility exist of a true sea monster as yet unknown to science?

In terms of hard scientific proof, given the reported frequency of visits by sea serpents, there is precious little evidence to show for the many sightings. Purportedly credible eyewitnesses include priests, naturalists, and Eagle Scouts. Many more times, however, eyewitnesses are individuals of doubtful veracity, opportunists who hope to profit from their reports in some way. There are no tissue samples, no bodies, no decent photos, and nothing more than what the courts would dismissively toss out as being merely hearsay evidence.

Round up the usual suspects and you'll find a list of real animals that *might* conceivably be mistaken for a sea serpent. These include the frilled shark, the conger eel, the oarfish, groups of whales or dolphins swimming in a line (think Olympic-style synchronized swimmers who just happen to be cetaceans), and giant squids. One of my favorite contenders for the sea serpent is the presumed-to-be-extinct Steller's sea cow, but more on that suspect later.

Take at look at Bruce Champagne's "A Preliminary Evaluation of a Study of the Morphology, Behavior, Autoecology, and Habitat of Large, Unidentified Marine Animals, Based on Recorded Field Observations" and you'll find a meticulously detailed listing of more than 1,200 reports of sea serpent sightings throughout the world. The more gullible—or perhaps more hopeful— researchers attribute reports to pocket populations of prehistoric plesiosaurs, mosasaurs, and other extinct marine reptiles, despite there being no concrete evidence to support such suppositions.

Myths of sea serpents have a long history in Norway, where obvious exaggerations tell of serpents who go about on both sea and land, live in caves, and plunder sheep when the mood strikes them. The cold and remote northern coastline of Scandinavia seems fraught with peril. Monsters lurk around every cove and dwell in the depths of every fiord. Originating in the mists of Norse mythology, St. Olaf himself claims to have killed a sea serpent in 1028 AD.

One particularly detailed, serious report was given to the British Admiralty by Captain Peter M'Quhoe of the HMS *Daedalus* between Africa's Cape of Good Hope and St. Helena. The incident occurred on August 6, 1848. The captain and crew say they were very close to the animal. They saw the head like a snake of an unknown creature rise about four feet from the water while its body was about sixty-feet long. It sounds like a giant squid, but it had a "mane" or possibly seaweed covering its head. That particular description occurs often for sightings of unknown beasts, often enough to make scientists wonder what it might actually be, assuming mysterious sea creatures haven't taken to sporting seaweed hats. The report prompted speculation, for the first time, that a relict plesiosaur population might have survived. This event is included in Jonathan Eyers's *Don't Shoot the Albatross!: Nautical Myths and Superstitions* (2011). Eyers mentions that Sir Richard Owens, a well-known English biologist, thought the animal was most likely an elephant seal, which has a maximum length of perhaps thirteen feet—and no mane, obviously.

Seafarers from practically every country with a coastline have given detailed reports of notorious sea serpents or monsters lurking in the sea, some of them said to be longer than the biggest ships of the day. It's relatively easy to dismiss stories of great scaly beasts with long manes that resemble seaweed, especially when such reports come from hundreds or even thousands of years ago. How, though, can we dismiss the reports in modern times, especially when such reports are widely corroborated by a large number of people? What are we to make of reports by road work crews and Eagle Scouts of the sea serpent that surfaced off San Francisco in the 1880s and again in the 1980s?

The San Francisco sea serpent has a fairly well-documented history. A *New York Times* article published on April 5, 1885, reads: "According to the statement of J.P. Allen, of the Bank of California, he and several residents of Alameda were standing on the deck of the ferry boat *Garden City* yesterday morning (March 27) at about 8:30 o'clock between Alameda and Goat Island (now called Yerba Buena Island), when a huge, black monster suddenly raised its head and neck above the water to a height of about ten feet displaying a mouth two feet wide filled with two rows of sharply pointed teeth, and after taking a curious glance at the passing steamer, plunged again into the water, at the same time elevating a sixty-foot long tail, with which it thrashed the water for some time, after which it made off in the direction of the Alameda baths, near where some fishing boats were anchored." The *Times* presents this feature not as hard news but as a "fish tale," humorously related and written tongue-in-cheek. Since then, the original sighting has been referenced as giving credence to a second and third appearance of a similar monster some one hundred years later.

The fact that the next report was given on Halloween 1983 ought to be viewed as a bit coincidental. The only more suspicious date for a sighting might be April Fool's Day. A road crew of five men was working on a section of highway when one of them spotted a long serpent heading through the surf toward where they were standing on the cliffs above Stinson Beach. I lived in San Francisco for a number of years in the 1970s and can tell you from first-hand experience that this area is remote, rocky, cold, and a very good setting for spotting a monster—or at least for making that claim. It's often shrouded in mist.

One of the road crew supposedly called to another member on the walkie-talkie, telling him to check out what was swimming in the water about a hundred yards offshore. The only place this apocryphal story turns up is in sensationalist literature, where it is repeated over and over again, even down to the colorful quotations given by the road crew. The serpent was said to be "boogying," for example. Marlene Martin was the Caltran safety engineer on duty. "It shocked the hell out of me," she said. "That thing's so big he deserves front-page coverage." The article, "Sea Serpent Captured on Paper," was written by Steve Rubenstein in the *San Francisco Chronicle* and (ostensibly) published on November 3, 1983. If something like that actually happened, you'd think it would be relatively easy to verify. It isn't.

The next time we hear about the San Francisco sea serpent is on February 5, 1985, when the brothers Clark report sitting quietly in their car sipping coffee and casually watching the bay. First, the twins see a group of sea lions, not at all unusual in the area. If you've ever visited PIER 39, you can vouch for that. If you haven't, you can check it out on the live webcam that shows them lounging around very clearly. Something was chasing them. Bob was "able to make eye contact," he writes, with one of the sea lions being chased. He could "see the fear of death in its eyes." Although he estimated the sea serpent as still being 25 yards away, he was able to describe it with uncanny detail: it was "black and slimy" with "maybe thirty feet of neck."

Much has been made, most often by the brothers themselves, of the fact that the Clarks were Eagle Scouts, and while that accomplishment is certainly laudable, it does *not* guarantee the authenticity of their story. They managed to sight the many-humped monster on several subsequent occasions and took a few terribly fuzzy photos of the beast. It had "hexagonal scales that fit next to each other, rather than overlapping." The black color on top "changed to a mossy green and then to a grassy green and ultimately to a yellow-green...". In a publication titled *Dracontology, Special No. 1, Being an Examination of Unknown Aquatic Animals*, Bill and Bob Clark spend an entire page asking the reader in various ways what they could possibly have to gain by creating such a story. The answer seems obvious—plenty! They have practically turned their sea serpent "sightings" into a cottage industry.

They saw it next on February 28, and then on March 1, 1985, and again on December 22 and 23, 1986, and *again* on January 24, February 25, and March 1, 1987. The monster appeared to the Clarks so often, you'd almost think it had developed something of a crush on the brothers. The boys were, just like most observers of such phenomena, notoriously, even ridiculously, bad at photography. Their images, according to "experts," show what are described as "a tilted fuzzy smudge" or "two smudges" and "a discontinuous line." One photo, it was suggested, might have been "a diving bird."

People concoct hoaxes for many reasons, not all of which are obvious to outside observers. Given the excruciatingly detailed and illustrated reports and the notarized statements provided by the brothers, it seems pretty clear that they are spinning quite an elaborate yarn. It's the kind of story that has been sometimes recanted in deathbed confessions by other hoaxers.

There *was* an enormous sea creature in the same area, however, at almost the same time. Humphrey the humpback whale, a famous cetacean who spawned a children's book and many t-shirts, did swim up from the bay and into the Sacramento River which he followed for a number of miles in November 1985. He finally left the vicinity on November 4. Humphrey was estimated to weigh perhaps forty tons. He got within 27 miles of the state capital, Sacramento. In 1990, he returned to the area and may have got himself stuck twice in the shallows near Candlestick Park. Some people said they saw him there. He also took a well-documented tour around Alcatraz and PIER 39, just like many tourists do on the harbor ferry.

Most people assume that when sea serpents do appear, they are simply strange marine animals, creatures unfamiliar to those who observe them. Let's take a look at some popular contenders.

If you're looking for a true serpent of the sea, the conger eel comes to mind. This homely guy, sometimes used in sushi, can grow to be twenty-feet long. People are very poor judges of size and distance on the open sea where they are few or no visible landmarks to use for comparison. One record-breaker conger eel caught off Plymouth in Devon, England, in 2015 weighed close to 160 pounds and was "longer than some buses." "What a beast!" the fisherman who caught it posted on Facebook. The conger can and has attacked humans. Writing for the *Irish News* on July 13, 2013, reporter Edna Dowling details one such confrontation in an article titled "Diver 'felt like a rag doll' in frenzied conger eel attack."

Diver Jimmy Griffen says in the article that he feels "extremely lucky to be alive" after the eel bit off a large chunk of his face. A gruesome photo of Jimmy in the hospital, minus a good bit of his cheek, accompanies the report. He was taking part in a diving expedition in Killarney in County Galway, Ireland, and was a diver with twenty years experience. "It gripped on to my face and threw me about violently. It was biting, pulling, and twisting on my face," he said. "When it finally let go, I could see that it was a conger eel swimming away from me, bigger than myself, so over six feet in length. It was about the width of a human thigh, so it was very strong." It would give a person shivers to imagine what a *twenty*-foot long conger like the one hooked off the coast of Devon in 2015 might have done to Jimmy's face, if this kind of damage could be inflicted by one merely six feet in length.

Another long, strange, mostly unknown sea serpent look-alike is the oarfish. Some of the old pen-and-ink drawings of sea serpents, in fact, look like textbook illustrations of an oarfish. It can reach lengths even greater than the conger eel, up to thirty-six feet and more. The *BBC Earth News* on February 8, 2010, posted a video of a solitary oarfish swimming at great depth in the Gulf of Mexico. Reporter Jody Bourton writes that this variety of fish may reach upper limits of 17 meters, or slightly more than 55' 9" long! They are the largest of any bony fish and live in temperate or tropic seas. These creatures are rarely seen by people unless the oarfish are dead or dying when they wash up on shore. Small dorsal fins, some four hundred of them, stand up in a line along the backbone of the fish, definitely lending it that *je ne sais quoi* so favored by illustrators of sea serpents. Those fins on its head are many times longer than those decorating the spine. It isn't good to eat, having the consistency of fish-flavored jello, so it is not commercially viable. The oarfish is silvery in color with black dots and squiggly lines but has no scales. Some of them have long, red streamers about their faces. Because they float and drift at the surface in their death throes and swim in an undulating, up-and-down movement, ichthyologists believe that many sightings of sea serpents were most likely harmless oarfish.

Frilled sharks are sometimes suggested as being sea serpent look-alikes. They are found in the Atlantic and Pacific Oceans, but despite a fearsome, prehistoric appearance, this shark only reaches lengths of about six feet. Most sea serpents are reported to be in the hundred-foot long neighborhood. The frilled shark has a body like an eel and lunges at its prey like a snake. Some suggest that perhaps an extinct relative, *Chlamydoselachus*, of the frilled shark may have survived, a much *bigger* kind of frilled shark. Again, we have no evidence of this.

Finally, let's consider my favorite contender for the elusive sea serpent, the Stellar's sea cow. Georg Wilhelm Stellar wrote about and sketched from a relict population of some 1,500 of the animals while he was shipwrecked near Bering Island in the Pacific Northwest. From the time Europeans discovered this poor animal, it took them just twenty-seven years to kill and eat every one of them. It was extinct by 1738. This gentle giant, a placid relative of the manatee and dugong, was about thirty-feet long and weighed perhaps ten tons. Its head did look much like the frequently mentioned "horse" head of the sea serpent. Could a few have managed to elude hunters and breed among the most isolated coasts among Russia's Commander Islands? I'd very much like to think so, but without some proof (and I'm not talking about fuzzy photos taken by Eagle Scouts), sea serpents continue to remain legendary if ubiquitous.

If you're looking for a *real* sea monster, look no further than the leopard seal, *Hydrurga leptonyx*. The elephant seal is larger, but this huge predator has sharp, inch-long canine teeth, can weigh upward of 1,400 pounds, and reach between twelve and thirteen feet in length. They swim as fast as great white sharks, 25–35 miles an hour. On land, they are every bit as fast as humans, and we are at a distinct disadvantage on the ice and snow. They range as far as the southern coasts of Australia and South America. Described as fierce and formidable, they feed on warm-blooded prey. Imagine a very big, streamlined, muscular seal with the appetite and hunting instincts of a tiger. Along with killer whales, they are Antarctica's apex predators. They will eat other seals, penguins, and will go after humans if the opportunity presents itself.

A "fierce, handsome brute," wrote Frank Worsley, Sir Ernest Shackleton's skipper on the famous 1914 *Endurance* expedition, according to Kim Heacox in *National Geographic*, November 2006. "Thomas Orde-Lees was skiing across sea ice when a leopard seal emerged from between two floes and lunged after him in bold, snakelike movements. Orde-Lees managed to keep ahead, kicking and gliding, until the seal dived into an open lane of water and tracked him from below—following his shadow—to pop up ahead." Garth Wood, a Scottish explorer, was also targeted when in 1985 a leopard seal grabbed him by the leg and tried to drag him into the water where he wouldn't have stood a chance. Only repeated kicks to the head by his companions wearing spiked boots managed to discourage the fearsome predator.

"Leopard Seal Kills Scientist in Antarctica," reads the title of an article written by James Owen in England in *National Geographic News*, August 6, 2003. Researcher Kirsty Brown, twenty-eight, was snorkeling when she was seized and carried some two-hundred feet under water. She had multiple, piercing head wounds. Owen interviewed Ian Boyd, director of the Sea Mammal Research Unit in Scotland, who said, "As with any top predator, like tigers and polar bears, leopard seals are charismatic creatures, but that isn't sufficient to justify the extremely high costs of working with them." Boyd believes there are three reasons that would explain why the diver was targeted. It might be defensive, assuming the seal believed itself to be in danger. It might mistake the diver for a seal, which it would be hunting. "A third possibility is more worrying," Boyd cautions.

"Leopard seals have been known to stalk people, and it's possible an animal could attack a person while knowing exactly what it's attacking." Leopard seals are regarded as inquisitive animals, but Boyd says divers should be wary of letting themselves become objects of curiosity. He said, "I think this inquisitiveness toward humans is to do with sizing them up as potential prey."

That's precisely what happened to biologist Lisa Kelly in Antarctica. A thirteen-foot long leopard seal repeatedly bumped up against her. Every time Lisa moved, the animal did, too. Watch her YouTube post (September 8, 2014) of the entire filmed encounter to determine whether you think the incident is benign curiosity on the part of the seal or prey-seeking, human-stalking behavior. Lisa professes she is now "absolutely in love" with the leopard seal. It even followed her back to the research vessel and from there back to land. If she continues interacting with charismatic leopard seals in this way, she may even find herself invited over for dinner by a leopard seal one day—as the main course.

CHAPTER NINE

The Gloucester Sea Serpent

"Is that it, Andy?" My little sister had been hoping for a glimpse of the sea serpent ever since last year when it seemed like half the town encountered it each and every time they approached the sea. Her china blue eyes sparkled. Loose golden curls blew around her head.

"No, Tess, that's not it. Look closer. That's a school of porpoises." I hated to disappoint her, but I wasn't going to lie to her, either. "See how they frolic up and down? Those humps aren't connected to each other. They are the backs of the animals."

"Oh, yes. I see now, Andy." She was dejected but undiscouraged. Tess continued to scan the horizon hopefully, just as she did every summer day when she and I walked along the rocky footpaths leading around Brace Cove past Niles Pond, our favorite place to ice skate every winter, and out to the tip of Eastern Point, from whence we could see the entire expanse of the Atlantic Ocean stretched out before us.

Gloucester was becoming famous these days, and not just for fishing, either, although it was the industry that supported most of the residents in town. The year was 1818, and it seemed the ocean would never run out of the cod that sustained the fishermen's trade. This was the year mackerel had been elevated to the status of edible fish and not merely something to use for bait. The vessel *President* under Captain Simeon Burnham made a trip out of the Inner Harbor for mackerel this year—the town was abuzz with the possibilities the change would bring about. The town was also buzzing with talk of the sea serpent's return.

"Look, look there!" Tess commanded, as imperious as any royal princess.

Reluctantly, my gaze followed the direction of her arm and chubby index finger. We'd been on the lookout for weeks and seemed to be the only residents in Gloucester destined not to see the monster.

A squarish dark head shaped something like a horse's but on a longer neck rose up out of the foam, dripping. Behind it bobbed a series of rounded segments. The closest I can come to describing them is to say they reminded me of the segments of an earthworm or caterpillar, and it moved in the same curious way an inchworm moves, except that this creature was enormous. I'd estimate its length at perhaps a hundred feet, and I'm a pretty fair judge of measurements.

"Yes!" I told my sister, lifting and twirling her in a little circle around me before setting her back down to again watch the serpent. "Surely that's it, Tess! What else could it be?"

It was dark and shiny and impossible to miss. There was no mistaking the mysterious, magnificent creature before us. Huzzah! At last, we, too, had seen the Gloucester sea serpent!

One sea serpent stands—or swims— above the rest. It is widely known as the Gloucester sea serpent, but it was seen in multiple places. Many people from all walks of life observed the creature or its progeny multiple times over hundreds of years. While it might not have been a sea serpent, there can be little doubt that people along the Massachusetts coastline were seeing *something* strange in the water, something large and something alive.

From the Celebrate Boston website, comes this information: "Several sightings by vessels from Gloucester, Massachusetts, took place during 1817–1818. A Gloucester sea captain attested to being attacked by a giant sea serpent. The captain took a couple of shots at the animal with his pistol, to no avail. Collective witness testimony makes it likely that what the sailors observed was a giant squid. The attack is not unlike Captain Nemo's battle in Jules Verne's *20,000 Leagues Under the Sea*." If you are tempted to think the witnesses were basing their testimony on that famous book, don't be. They hadn't read the book because it wasn't published until 1870.

An article in the May 16, 1818, *Essex Patriot* describes this memorable event:

> AFFADAVIT (sic)
>
> I, Joseph Woodward, Master of the Sch. *Armament* of Hingham, on my passage from Penobscot to Hingham, on Saturday last, at

2 o'clock, P.M. *Agementicus* bearing W.N.W. ten leagues distance, discovered some thing on the surface of the water apparently about the size of a ship's long boat. Supposing it to be the wreck of some vessel, I made towards it; and on approaching it, to my surprise and that of my crew, discovered it to be a monstrous Sea Serpent—as we approached him, he threw himself into a coil & darted himself forward with amazing velocity—the wind being ahead, it became necessary to stand on the other tack, and as we approached him again, he threw himself into a coil as before, and came across our bows at not more than 60 feet distance.

Having a gun charged with a ball and shot, I discharged the contents of it at his head. The ball and shot were distinctly heard to strike him and rebound as though fired against a rock— he, however, shook his head and tail most terribly—he again threw himself into a coil and came towards us with his mouth wide open. In the meantime, I had charged my gun again and intended to have discharged the contents of it into his mouth; but he came so near us, I was fearful of the consequences, and withheld it—he came close under the bows of the [schooner] and had she not been kept away must have came on board of us—he sunk down under the vessel, his head a considerable distance on one side the vessel and his tail the other—he played around us about five hours—I and my crew had probably the best opportunity of seeing him that has occurred—I judge him to be, at the least twice the length of my [schooner] say one hundred and thirty feet—his head was about the size of a ship's long boat, say fourteen feet—his body, below the neck, at least six feet in diameter—his head was large in proportion to his body—his tail was formed like a squid's—his body was of a dark color and resembled the joints of a shark's back bone—his gills were about twelve feet from the end of his head, and his whole appearance was most terrific.

His manner of throwing himself into a coil appeared to be done by contacting his body in a number of places in perpendicular directions, and placing his tail so as to throw himself forward with great force—he could contact and throw himself in any direction with apparently the greatest ease and most astonishing celerity.

Joseph Woodward

Hingham, May 12, 1818

Having read the above statement of Capt. Woodward, we certify to the correctness of it.

Peter Holmes

John Mayo

Plymouth, ss. May 12

Personally appeared, Joseph Woodward, Peter Holmes, & John Mayo and made oath, that the above statement is just and true— before me.

Jotham Lincoln Jr.

Just. Peace.

The captain of the schooner *Armament* is clearly in earnest. Something was observed, fired upon, and hit. The tone is sober yet full of wonder at this unforgettable encounter with a mysterious sea beast. Was it a giant squid? "Its tail was formed like a squid," Woodward notes. It was long, flexible, and moved rapidly. Perhaps it was, in fact, a squid, although why the captain, who was presumably familiar enough with the appearance of a squid to know what its "tail" looked like, didn't think so is puzzling.

It wasn't the first sighting of the serpent by Europeans. John Josselyn wrote in 1638:

They told me of a sea serpent, or snake, that lay coiled up like a cable upon the rock at Cape Ann; a boat passing by with English on board, and two Indians, they would have shot the serpent, but the Indians dissuaded them, saying that if he were not killed outright, they would all be in danger of their lives...

A report by Obadiah Turner, either less thoroughly edited than Josselyn's account or written by a poorer speller, made in 1641, just twenty-one years after the *Mayflower* arrived, provided a report of the Gloucester sea serpent as it appeared off the coast of Lynn, Massachusetts:

Some being on ye great beache gathering of clams and seaweed which had been cast thereon by ye mightie storm did spy a most wonderful serpent a shorte way off from ye shore. He was big round in ye thickest part as a wine pipe; and they do affirm that he was fifteen fathoms [90 feet] or more in length. A most wonderful tale. But ye witnesses be credible, and it would be of no account to them to tell an untrue tale. Wee have likewise heard yet Cape Ann ye people have seene a monster like unto this, which did there come out of ye land much to ye terror of them eyt did see him.

I'm not sure about the thickness of a "wine pipe," but let's assume what people saw on "ye great beache" was pretty wide and very long. Since Cape Ann is located just up the North Shore from Lynn and was said to have been visited by the serpent several years previously, it could well be a sea creature hanging about in familiar territory.

The New England Linnaean Society, so-called in honor of Carl Linnaeus who standardized a system for naming fauna, created a group tasked with finding answers about the mysterious beast on August 18, 1817. These men called the animal *Scoliophis atlanticus*, or Atlantic humped snake. The team came in for a good deal of ridicule when they next made the claim that a deformed black snake (it had humps on its back) found in Loblolly Cove in Rockport was actually a baby sea serpent. The committee concluded, despite a lack of any evidence tying the two creatures together: "It is worthy of remark, that nearly all the circumstances with regard to the appearance of the Great Serpent, stated by the deponents, as facts, agree with the structure of Scoliophis... Supposing that the species of the two serpents is the same, it is not improbable that one is the progeny of the other." Maybe it wasn't *impossible*, but it was highly *improbable*.

The next year, the serpent came back to visit the waters off Portland, Maine, and Cape Ann, Salem, and Ipswich, Massachusetts. A sea serpent was frequently seen at Nantasket Beach, Massachusetts, as was reported in *The Boston Globe* in 1830, 1875, 1905, 1913, 1921, 1926, and 1937. Nantasket is located in the town of Hull on a small peninsula just south of Boston. Some of these accounts included details about two very large, round eyes about two feet apart, a trait characteristic of the giant squid. The most recent sighting happened in 1962, off the coast of Marshfield, Massachusetts, which is a bit south of Hull.

The South American fresh water anaconda approaches forty feet in length at the outside and is as thick as a good-sized tree trunk. Could a salt water variety as large as the anaconda have been responsible for the reports of the Gloucester sea serpent? There's no evidence to support such a claim, but it's a hypothesis put forth with some merit. Giant, yet-to-be-discovered sea snakes? That's no more outlandish than the megamouth shark or the coelecanth, is it?

The best explanation I've come across to date is the one given by Elizabeth Fama writing for Tor.com on August 16, 2012, in an article called "Debunking the Great New England Sea Serpent."

People interviewed about the monster, she notes, were always careful to rule out possible explanations. It wasn't porpoises swimming in a line, wasn't long ropes of seaweed, wasn't a school of fish. Fama says the answer to the enigma may be found in the very words of the accounts themselves:

> "...like a string of gallon kegs 100-feet long."

> "He...resembles a string of buoys on a net rope, as is set in the water to catch herring."

> "The back was composed of bunches about the size of a flour barrel, which were apparently three feet apart; they appeared to be fixed but might be occasioned by the motion of the animal, and looked like a string of casks or barrels tied together..."

Fama urges readers to consider the logic of the duck. Remember that one? If it *looks* like a duck... Well, she believes, maybe what witnesses were seeing actually *were* barrels, casks, buoys, or kegs strung together by nets attached to a humpback whale! Brilliant. That doesn't rule out other creatures of the deep like the oarfish or giant squid being spotted off Massachusetts on occasion, but it certainly does explain the majority of the descriptions of the historic beast.

This past summer, I took a whale watching boat out of Gloucester. Within a few minutes, the boat was surrounded on all sides by humpbacks. There were bubble nets created as they hunted and feasted in the plankton-rich waters . They breached and slapped their tails. In a very short time, it became impossible to count all of the whales. I stopped at fifty. They are common in these waters around the Stellwagen Banks National Marine Sanctuary, can and do become trapped in nets, which is why the use of drift nets was discontinued in modern times, and follow the same routes from breeding grounds in the Caribbean to feeding grounds in the North Atlantic. I believe Elizabth Fama's idea is a genuinely plausible one.

"The beast's heyday," she says, "was between 1817 and 1819, when hundreds of people saw it in the Gulf of Maine... Once, 'a crowd of witnesses exceeding two hundred' watched it, at various angles and altitudes from shore, for three-and-a-quarter hours. In the summer of 1817, the animal lingered so long and often in Gloucester that, 'Almost every individual in town, both great and small, had been gratified at a great or less distance with a sight of him.' Families saw it; sailors; captains; whalers; and even a couple of naturalists saw it. Men shot at it with rifles and tried to impale it with harpoons. It seemed impervious.'" Maybe they were shooting the barrels?

She goes on to include verbatim descriptions provided by witnesses, amended with her conclusions in parentheses, following those accounts:

> "[He appeared in] exactly the season when the first setting of mackerel occurs in our bay." [Whales eat schooling fish like herring and mackerel.]

> "...claimed he'd seen a sea serpent about two leagues from Cape Ann battling a large humpback whale." [Proximity of a whale to the serpent.]

> "At this time [the creature] moved more rapidly, causing a white foam under the chin, and a long wake, and his protuberances had a more uniform appearance." [The foam suggests something is pulling the object, and the strand of kegs elongates when towed.]

> "...the times he kept under the water was on the average of eight minutes." [Like a whale.]

"In the early 19th century," Fama explained, "a purse seine net would likely have had cedar or cork floats. But after a bit of research I found that small wooden casks were used as buoys and as floats for fish nets in Newfoundland and Norway in the 1800s."

She came to this insightful conclusion after viewing a program called "Saving Valentina" posted on YouTube on June 13, 2011, about a humpback whale that had been ensnared in fishing nets. Who knows? Maybe the Gloucester sea serpent was like Valentina, a victim of nets. That's the best, most original, and believable answer I've found to date.

Note: To anyone wishing a deeper immersion in the subject, I recommend *The Great New England Sea Serpent: An Account of Unknown Creatures Sighted by Many Respectable Persons Between 1638 and the Present Day,* by J.P. O'Neill (1999). Another quite detailed, very readable account of the Gloucester sea serpent appears in the fascinating book *Monsters of the Sea* by one of my favorite authors, Richard Ellis (2006). Both volumes are filled with many anecdotes about the Gloucester sea serpent and reports taken from the witnesses describing its appearances. These books supplement the text with a rich historical context and interesting illustrations.

CHAPTER TEN

The Globster

"What is it?"

"How should *I* know?"

"Touch it."

"I ain't touching it. *You* touch it!"

The two boys, fondly known as "sand rats" by the older folks in town, regularly prowled the beach to get the first look at anything interesting that might have washed ashore overnight. They rode their bikes down to the beach early to get there before anyone else was up. Usually. if they found anything at all, it was nothing more than a primitive horseshoe crab, a strange animal that looked like it could have been on a first-name basis with the trilobite, or the strange lumpy shape of a sea cucumber or on one exciting occasion the dark, worn tooth of a shark. Shark skeletons are cartilage, not bone, so teeth are all they leave behind to mark their passing. Today was different. A truly enormous shape dominated this stony, protected cove, the one they knew as well as their own backyards.

"Y'think we oughta tell somebody?"

"Yeah."

Neither boy could tear his eyes away from the giant, grayish lump of flesh. It appeared to be covered all over with small, white hairs. There were no eyes, no mouth, no nostrils, no legs or fins visible. Still, it was definitely *something*, something big and important, that was for sure.

"Let's go!" one finally yelled to the other.

"Okay!"

They rode like the very Devil was after them.

Within an hour, the usually quiet beach was swarming with crowds of excited people. This would surely put their little seaside town on the map! Reporters from big-city newspapers descended in droves by lunchtime. They brought photographers with them. The two boys, the ones who made the discovery, were interviewed over and over again. Their parents, dressed in their Sunday best, stood beaming proudly behind their sons. The mayor, puffed up with self-importance, held a press conference. Although he knew next to nothing about the mysterious creature under scrutiny, that didn't stop him sharing his opinions. By the time evening editions went to press, headlines across the whole state would be screaming the news—*Globster*!

Animals on land die, and there is a carcass to discover and examine. Sometimes, it proves to be a mangy coyote and not the Chupacabra, but at least that determination can be scientifically verified. When sea animals die, considerable time may pass before the carcass, or what is left of it, washes up on some sandy shore. When it does, people flock to the scene to view the puzzling remains of the latest unknown creature of the deep. Is it a giant specimen of a hitherto unknown species of squid, a sea serpent, a distant cousin of the Loch Ness monster? In the past, people were left to puzzle out the source of the blob on the beach with nothing more than speculation. Before the days of photography, sketches were made. Once photography became possible, the remains were documented on film and the strange images published in newspapers. Occasionally, tissue samples were even preserved. Today, we can do more.

Now that DNA analysis is available, it is finally possible to identify the great majority of the large, intriguing lumps. If they can't be identified, it's because viable strands of DNA couldn't be extracted from the rotting flesh, not because the creature was one of an unknown species. Most turn out to be nothing more mysterious than a decomposed whale or large hunk of whale blubber, basking shark, squid, oarfish, sea lion, or some other large marine animal. Records dating back to 1648 record details of dozens of these carcasses. So far, none of them has proved to be the remnants of an unknown sea creature. Some, however, are certainly more interesting than others.

Stacy Conradt, writing for Mental Floss in January 2008, says, "Globsters have the makings of a terrible (but great) horror movie. What is a globster? So glad you asked...it's kind of hard to define. Basically, it's a blobby-looking, unidentifiable carcass that has washed up on a shore somewhere. Some globsters have bones, some don't. Sometime they have tentacles, flippers, and eyes, sometimes they don't." Let's take a closer look at some of the most famous of the globsters, also called blobs.

The St. Augustine Monster was discovered in 1896 by some kids riding by the beach on their bicycles. It generated a frenzy of popular interest. Ultimately, its pinkish skin and the presence of long tentacles, many of them severed and scattered over the sand, created a consensus that the monster was some species of giant squid. This was at a time before the existence of *Architeuthis dux* was recognized by scientists, something that didn't happen until 1925. Fortunately, a piece of the St. Augustine Monster was preserved and kept at the Smithsonian Museum.

Cell biologist Dr. Joseph F. Gennaro Jr., working at the University of Florida, compared the connective tissue of the St. Augustine carcass to control specimens from known octopus and squid species. His findings appeared in the March 1971 issue of *Natural History*. He wrote:

> We decided at once, and beyond any doubt, that the sample was not whale blubber. Further, the connective tissue pattern was that of broad bands in the plane of the section with equally broad bands arranged perpendicularly, a structure similar to, if not identical with, that in my octopus sample. The evidence appears unmistakable that the St. Augustine sea monster was in fact an octopus, but the implications are fantastic. Even though the sea presents us from time to time with strange and astonishing phenomena, the idea of a gigantic octopus, with arms 75 to 100 feet in length and about 18 inches in diameter at the base—a total spread of some 200 feet—is difficult to comprehend.

How true. Not only was it difficult to comprehend, it turned out to be impossible to comprehend because it was misidentified.

Roy Mackel, a biologist at the University of Chicago and founding member of the International Society of Cryptozoology (founded in 1982, disbanded in 1998), did his own experiments which, not surprisingly, agreed with Gennaro's findings. It was definitely not blubber, he asserted. It was from *O. giantius*, an unknown species of octopus. In 1986, his tests showed that the sample was composed of "masses of virtually pure collagen" and did not have the "biochemical characteristics of invertebrate collagen, nor the collagen fiber arrangement of octopus mantle."

In 1994 when DNA testing became available and was done, it turned out that the actual animal that washed up on a Florida beach was, in fact, the remains of a sperm whale. Sidney Pierce and a team of biologists published the opinion that the majority of Globsters were whale remains in the April 1995 issue of the *Biological Bulletin*. In 2004, he conducted examinations on all available globster specimens using electron microscopes and molecular and DNA analysis. The conclusion was that all the samples they looked at came from different whale species.

No matter how much people might wish and hope that scientific evidence will eventually, definitively prove the existence of an enormous species of octopus (even if that discovery would be something of a double-edged sword—both fascinating and terrifying) or some other monster of the deep, we are left to conclude that the hope hasn't been realized to date. Still, those gigantic, amorphous globs continue to wash up on the world's beaches, and they continue to capture the public's imagination.

In 1960, a "huge" Globster covered with "fine hair" was found in Tasmania. "Nearly as big as a house!!" headlines screamed, assuming that the house in question was twenty-feet long and a couple of feet wide. It had no visible eyes or mouth, but the Tasmanian Monster was thought to be, perhaps, a new species of mammal. Eventually, tests made in 1981 of the collagen fibers proved it was a badly decomposed whale.

Japan, not to be outdone, lays claim to the Zuiyo-maru. It didn't wash up on a beach but was instead discovered by Japanese fisherman in 1997 aboard their boat of the same name. The Globster was pulled from the ocean after being caught in the net. The men aboard were instantly convinced they had discovered a relict plesiosaur, which they rather unoriginally dubbed Nessie, after Scotland's Loch Ness Monster. Samples of the creature were taken, along with photos. In the pictures, the decomposing, pinkish animal does look very much like the ancient plesiosaur. As far as unbiased, scientific observers are concerned, observers *not* among the crew of the *Zuiyo-maru* fishing vessel, it most closely resembles the rotting carcass of a basking shark.

Bermuda Blob the First was found in 1988, while Bermuda Blob the Second was found by fisherman Ted Tucker in 1995. It was a brilliant white color with five distinct "limbs." At the time, tests to determine its species were not made. It was believed to be related to *Chondrichthyes*, sharks or rays. It had no eyes, mouth, or features to distinguish it. The mystery didn't last very long.

Sidney Pierce and his team performed their tests on the tissue and the results were published in the April 2004 of the *Biological Bulletin*. The conclusion? Globsters are whale based. For example, the Chilean Blob is the almost completely decomposed remains of the blubber layer of a sperm whale. "This identification is the same as those we have obtained before from other relics such as the so-called giant octopus of St. Augustine (Florida), the Tasmanian West Coast Monster, the two Bermuda Blobs, and the Nantucket Blob."

Whales die, and blubber floats. Globsters are usually no more complicated than that. Still, if something huge and unknown washes up on the beach, you can be sure it will draw a crowd.

CHAPTER ELEVEN

Three Notorious Globsters: Gambo, Trunko, and the Stronstay Beast

Three creatures washed up on shore like the other Globsters, yet these three were slightly more mysterious than the rest. Each one of them has fascinating history. They've been dubbed Gambo (1983), Trunko (1924), and the Stronstay Beast (1808).

Gambo showed up off the coast of Gambia on Bungalow Beach one morning and caught the attention of a fifteen-year-old boy named Owen Burnham. No photos were taken of it, but Owen had an interest in the natural world and happened to be a decent young artist. He carefully sketched the creature before word of its existence spread. The skin was oily and brown. It had been attacked by a predator, an attack that apparently killed it. Internal organs were exposed and one rear appendage had been chewed off. Burnham's sketch shows the fifteen-foot long animal with a head distinctly dolphin-like but with a body much wider than a dolphin. The sketch looks like something of a dolphin-crocodile hybrid. There are four short flipper-like appendages, one mostly chewed off, and a long tail. In an area where such a bounty given up from the sea meant the prospect of cash, Gambo's head was severed and sold to a tourist, who has yet to come forward with it for further testing. Unless or until that happens, all that remains of Gambo are the sketches and the story.

Trunko is the only one of the well-known Globsters observed to be alive and swimming before it died. That would seem to invalidate the usual "whale blubber" answer. In the waters off Margate,

South Africa, on October 25, 1924, Trunko was seen for several hours defending itself with its powerful tail from an attack by two killer whales. Days later, its body washed ashore. The name, Trunko, comes from the Globster's long, protruding trunk. The animal was covered, as are many Globsters, with white hair. Four photos were made but remained largely uncirculated until 2010. One shows a man poking its long trunk, said by witnesses to resemble an elephant's trunk, with a stick, an ignominious end to what was reputed to be a mighty warrior. When I initially learned about Trunko and his struggle to survive the orca attack, I assumed it was an elephant seal, possibly an albino.

Speaking of ignominious, when Sea World first opened in San Diego, I saw an elephant seal perform in a show there. He was enormous. He had a tiny guitar on a strap placed around his neck and wore a miniature Mexican sombrero. They called him Go-Go. It was sad to see him treated that way, so undignified, but the 1960s were altogether different in so very many ways from today.

The southern elephant seal, *Mirounga leonina*, is the bigger of the two species of elephant seals. This animal lives in the frigid Antarctic waters where fish and squid are plentiful. They come ashore to breed but otherwise remain in the water. Animal species do produce the occasional albino. Such albinos are often easy, visible targets for predators. Could Trunco have been an albino elephant seal? I was becoming convinced, even though there was no clear evidence of this, but it was just one of several theories. At the time, some thought it was polar bear because of the white fur. "Fish Like a Polar Bear" was the title of an article about Trunko in the December 27, 1924, edition of *London's Daily Mail*. Polar bears are not found at the South Pole, however, making that hypothesis highly unlikely, nor do polar bears have trunks.

After being subjected to modern analysis, the four photos taken of Trunko make it appear more probable that the creature was not alive, and never alive, but just another blob of whale blubber. The "white hairs" were probably collagen fibers. The animal flesh was most likely being tossed about by the orcas as they ate from it. It might have looked alive from the shore, but it probably wasn't. Without more proof, Trunco's true story will remain mysterious, but it was likely the same prosaic type of Globster as the others.

The final animal, known as the Stronsay Beast, can also be reasonably explained. It was sighted in 1808 off the Orkney Islands in Scotland. While fishing, John Peace saw a strange carcass on the rocks offshore being picked at by sea birds. Ten days later it was washed ashore by strong winds where it could be more closely examined. A local, George Shearer, took careful measurements and described the beast as "serpentine." It was fifty-five feet from nose to tail, with a neck that was ten-feet-three-inches long. Six stubby "limbs" protruded from beneath the body. It had "smooth as velvet," gray, "elastic" skin. Its head was about the size of a sheep's with eyes as large as those of seal. Bristled hair grew along the spine and down the tail.

The Natural History Society in Edinburgh called the hitherto unknown animal *Halsydrus Pontoppidani* after the Norwegian bishop who collected tales of sea monsters in the eighteenth century. Naturalist Sir Everard Home's curiosity was piqued by reports of the Stronsay Beast. He traveled personally to Stronsay to examine what remained of the carcass. After doing so, he concluded that the thing was nothing more than a decomposing basking shark, the second largest of the sharks after the whale shark, an animal relatively common in the Orkney Islands area at that time. Sir Edward found vertebrae on the body of the beast that were identical to those of the basking shark. The so-called limbs were what remained of its fins. The shark's large lower jaw is attached by only a small piece of skin. Once that rots away and falls off, the little skull alone remains. The carcass was probably that of one of the big sharks from the northern seas, now nearing extinction. Looking at the two 1808 drawings, similarities to the photo of the *Zuiyo-maru* monster are apparent.

This animal was truly unusual because of the longest basking shark on record is about forty feet, while the Stronsay Monster was measured at fifty-five feet. In that sense, it really was a monster.

CHAPTER TWELVE

The Beast of Gévaudan

Henri Michaux, aged ten, was proud of his newly prominent role, now that his older brother had finally married and moved out of the house. He was the man of he family. He pulled his shoulders back, drew himself up to full height, which admittedly wasn't very tall, and strode through the village. He didn't bother to greet any children younger than himself, walked past them without even acknowledging their existence, and put on a serious face. He had taken on the important work of tending to his family's small flock of sheep. Wool was the commodity that kept them alive through the lengthy, freezing months of winter. His older brother sheared the sheep, he carded the wool, his mother and younger sisters spun it, and then it was used to knit and sell warm shawls, socks, scarves, mittens, hats, and sweaters. Sheep were the foundation upon which his family's survival rested.

"Bonjour, monsieur le boucher," he called as he passed, nodding importantly to the butcher who stood outside his shop.

"Soyez prudent aujourd'hui," the butcher cautioned. Be careful today. Danger lurked in the forested mountains nearby.

Something was wrong. He knew it as soon as he caught sight of the milling, bleating sheep. One of them was down, a mother-to-be, her body swollen with the lamb that had been growing inside. Her throat had been ripped out, a gaping wash of red against the fresh green of the grass. Puzzled eyes stared from a mild face. A fly crawled across one of them. Wolf. The forest practically crawled with them. Sheep were lost, young calves, poultry, and even occasionally a child. He shuddered at the thought.

There was a low, ominous snarl from behind him. Before he could even turn, Bravoure flew through the air, launching himself fearlessly at the beast. The brave, loyal sheepdog savagely bit through the heavy fur around the creature's neck. It slung him off easily and advanced on the boy. Again, the fiercely determined dog threw himself at the beast. Wolf? What sort of wolf was red with stripes and a tail as long as that one? The paws were as large as those of a bear, and the short face was ursine as well. Its jaws were unnaturally huge and gaped wide, white fangs snapping at Bravoure who sustained a savage bite to his front leg. Henri had never seen its like. His scream echoed off the granite peaks.

Later than morning, Bravoure limped back to the village on three legs whimpering piteously, a fourth dragging uselessly. When the men of the town returned to the site of the attack, all that remained of young Henri would have easily fit into a hat box.

People in the mountains of southern France have not forgotten the Beast of Gévaudan. Ever since I first learned of the existence of this terrifying creature, it has fascinated me. It was real, and it killed many people and animals—killed them viciously. It's small wonder children for generations thereafter were warned never to skip off into the woods alone. There are several theories about what the beast was: wolf, wolf-dog hybrid, werewolf, escaped hyena, and perhaps even an extinct species of large, long-haired hyena, *Pacyhcrocuta breverostris,* indigenous to this region some 400,000 years ago, or the also-extinct bear/dog carnivore, *Amphicyon ingens*. Whatever it was, it was horrific. Some say it's the genesis of the "big bad wolf" of fairy tales, but tales of an evil wolf preying on the innocent predate the appearance of the Beast of Gévaudan by at least one hundred years. Between 1764 and 1767, in an area of about fifty square miles, it hunted in the forests and fields of France. Its range, once called Gévaudan, gave the beast its name. Now the place is known as Lozère, which is in the area of Haute-Loire, located in the Margeride Mountains in the southcentral region of France.

Attacks by wolves were not rare during this period in French history. Unburied bodies from large-scale wars attracted them in droves. As a result, wolves were accustomed to eating human flesh. In contrast to modern experts' claims that healthy wolves will not hunt humans, the reality at that time was quite different. Between 1590 and 1690, wolves in France killed at least 1,800 people, and expert Jean-Marc Moreceau reckons it closer to 9,000 in "The Wolf Threat in France from the Middle Ages to the Twentieth Century" (2014). Morceau isn't talking about only rabid

wolves, either. He includes this sad note made in the parish regis-
ter of La Chapelle-Thècle (Saône-et-Loire) on October 8, 1749:

> Marie, aged approximately 7 years, daughter of Jacques Prudent
> and his first wife, Tiennette Maroyer, was snatched from her
> doorway by a wolf and devoured in a field. Only her head, one
> arm and her stomach were found, and nothing besides. These
> pitiful remains were buried in the cemetery of this church the
> following day, fifth October, before my entire parish, who had
> gathered for Sunday Mass.

Morceau puts the number of wild wolves in France at the end
of the eighteenth century at 10,000 to 15,000. (In the British
Isles, wolves were mostly gone by the fourteenth century.) Written reports variously referred to the animals as "beast in wolf
form," "ferocious beast," "flesh-hungry beast," "ferocious wolf,"
"flesh-hungry wolf," "stag-hunting wolf," "man-snatching wolf,"
"voracious wolf," or simply "bad wolf." Morceau's concern is the
reintroduction of wolves into the French ecosystem without an
understanding of the historical background of wolf attacks in the
regions. He specifically comments upon the wolf as a serial killer:

> Recurrent spates of predatory wolf attacks hold a well-defined
> place in the French common memory: they were the work of
> what people at the time called "beasts," because the animals
> seemed to them so far removed anthropologically from the ordi-
> nary wolves that attacked only livestock. These series of attacks,
> each of which cast a shadow over a small region for several years,
> left their mark in people's minds as much for their terrible conse-
> quences as for the difficulty of exterminating the attackers.

While Morceau regrets the negative stereotype of the big, bad
wolf, he hopes to educate the public about what happened when
wolves and humans co-existed in close proximity in the past.

In 2012, Giovanni Todaro wrote a nearly 500-page book called
La Bestia del Gévaudan, Quando il serial killer è un animale ("When
the serial killer is an animal"). In it, he notes that attacks and
killings by wolves were extremely common until the nineteenth
century. People, especially the poor, had few effective weapons
available to them. Man was *not* the apex predator in the region,
the wolf was. Todaro stresses that perhaps tens of thousands of
human beings lost their lives to predation by wild wolves.

Apparently, writes Morceau, the infamous Beast of Gévaudan
was not the only lupine serial killer:

This is why, a century before the Gévaudan affair, the Beast of Gâtinais was just as notorious. The term 'beast' often surfaces in descriptions of the most dramatic affairs, and it is this same term that is used to refer to the most prolonged and deadly spates in this long history of attacks. Whenever the deaths and injuries could be counted in tens, and attacks continued for several months or even several years, 'beast' was the label that came to the fore in discourse.

What was it? Descriptions are intriguing and point to something beyond the norm. On the surface, it seems that the common wolf was the most likely culprit. Wolves were endemic in this region, as were fatal attacks by wolves. Based on eyewitness accounts, however, the beast doesn't sound like any regular wolf.

According to Abbé Pierre Porcher, who published a meticulously detailed scholarly tome titled *La Bête du Gévaudan* in 1889, people who saw it described the animal as being bigger than a wolf with a larger, squarish head, more like a dog-wolf hybrid. It had relatively small, pointed ears, and was perhaps the size of a calf. Its very long, bushy tail was often mentioned. The color of fur most often cited was reddish, possibly striped, with a long, black stripe running the length of the back. The trait that is perhaps most puzzling is the shape of its mouth, which seems to have been odder than that of any wolf's. Fiercely long fangs are frequently mentioned as being clearly visible. In drawings, the beast's mouth as depicted by artists who listened to the eyewitness accounts looks positively deformed, making misidentification of the animal almost a given.

The initial attack by the beast was unsuccessful. It charged a woman alone tending a heard of cattle. Luckily for her, a few bulls charged back with their horns lowered. The animal wasn't dissuaded and attempted another attack. Again, the bulls fended it off, at which time it ran into the forest. Not long afterward, a less-fortunate fourteen-year-old was killed. Janne Boulet became the first official victim. Targets of the beast tended to be mostly women and children who were alone and often tending animals in the fields. By the end of its singular reign of terror, more than one hundred people were dead and many others wounded badly.

This area of France is extremely harsh. Some residents say there are "nine months of winter and three months of hell," although that is a description more often applied to Madrid, in the mountainous regions surrounding Gévaudan. People had to labor very hard in order to make a living. If they didn't go out to work, they didn't eat. They existed on the precarious border between survival and death.

The killings continued and panic increased. Wolves generally go for the jugular when attacking. This animal tended to decapitate its many victims, an ironic foreshadowing of the terror of Madame Guillotine, bloody emblem of the French Revolution that would begin in 1789. The territory where the beast found its victims was only tenuously attached to France in 1764 to 1767, and the political situation there was unstable. Following one notorious attack, the attention of Versailles turned toward Gévaudan.

Even though the majority of victims were solitary women and children, easy to pick off, one particular attack was different. A group of children was targeted by the beast, and one among them was seized by the head and carried away. It happened on January, 12, 1765, in the village of Villeret. Jay Smith in *Monsters of the Gévaudan* (2012), describes the event that elevated twelve-year-old Jacques Portefaix to the status of a national hero overnight: "Instead of running away from the beast, he resolved to stand and fight. ... No adults were present to witness the confrontation, at least until the very end of the ordeal. All agreed that Portefaix had showed extraordinary courage."

After the beast dragged away the smallest and most vulnerable child in the crowd, "most of the others wanted to flee, for fear that the beast might next come after them." Here, Smith quotes from the *Courrier d'Avignon*, a newspaper of the day. Portefaix told them "they must pursue the beast, kill it if possible, or at least force it to let go of its prey." The coverage continues: "He put himself at the head of his comrades." The boys caught up to it, hitting it so hard on the hindquarters that it dropped the child in its jaws and ran away. The victim's wounds "were not mortal."

Even if the account was subsequently embellished in the press, the boy's heroism is noteworthy. The encounter came to the attention of King Louis XV. Young Jacques was paid 300 livres, while 350 livres more was given to be shared among the other boys. The king also directed that Portefaix be educated at the state's expense. He was to be groomed for a life in the military serving France. Louis then decreed that the French state would help find and kill the beast.

People didn't always run away in fear when they encountered the beast. There were many tales of incredible bravery in the face of savagery. This one is vividly described by Jay Smith. A mother, Jeanne Varlet, from Bessière was in the garden around her house with several of her children when the beast attacked, targeting the children and grabbing two of them in turn by their heads. She pulled first one and then another from its jaws by her children's' legs in an "awful tug-of-war" that lasted several minutes. As soon as she freed one child, the other was taken. The beast became agitated and frustrated, as did the mother, who eventually jumped on its back when trying to extricate her six-year-old son from its jaws. Falling off, she grabbed it by its "most sensitive parts of its body," according to the *Gazette d'France*.

The animal turned its attention from the six year old to the eighteen month old who was held in the arms of its sister, aged ten, and back again. Eventually, with the middle boy's head in its jaws, it sailed over the four-foot-tall wall around the garden. The mother never gave up. She chased it and finally her cries were heard by her teenaged son who came with their sheep dog and a sharp lance to take over the fight. The poor victim, badly mauled, did not survive the attack. He had lost his nose and upper lip, as well as sustaining terrible trauma, but the beast was driven off before it was able to do anything further.

There is a statue erected to commemorate the bravery of another young heroine, Marie-Jeanne Valet, who took on the beast. In 1995, sculptor Philippe Kaeppelin created a monument in Auvers to celebrate Valet's courage. The depiction was based on people's accounts of the beast's appearance. It's like no ordinary wolf. The jaws are massively outsized. At the time, Marie-Jeanne said it looked like a huge dog. She was walking home from a farm nearby, a journey that took her across a stream, when it ambushed her. Her sixteen-year-old sister was with her. The online *Atlas Obscura* describes it this way: "Valet was crossing between branches of a river through a small wooded area when she turned to discover the beast immediately behind her. As it reared up for an attack, the young woman plunged a homemade spear she had been carrying into its chest. Injured but not dead, the beast raised a paw to the injury, crying out loudly, and then rolled off into the waters of the river." It was bleeding but was proving hard to kill.

King Louis XV finally decided to hire professional huntsmen, a father-and-son team, to eradicate the beast once and for all. Jean Charles Marc Antoine Vaumesle d'Enneval and his son Jean-François were dispatched to Gévaudan. They came to Clermont-Ferrand in February 1765 and brought along eight trained bloodhounds. The father-and-son pair killed many wolves in the vicinity but attacks continued, so in June, they were replaced by Francois Antoine. In September, Antoine killed a wolf that was nearly three feet tall at the shoulder, over five-and-a-half feet long, and weighed about 130 pounds. Moreover, victims identified the body of the wolf based upon scars it bore from its confrontations with those who had managed to fend it off. The wolf was stuffed, sent to Versailles, and a large reward paid to the hunter, yet the attacks continued. At least a dozen more followed in quick succession.

The end of the terror came when local man Jean Chastel shot a similar beast during a hunt organized by a local nobleman, the Marquis d'Apcher, on June 19, 1767. Accounts of Chastel's success, Jay Smith points out in *Monsters of the Gévaudan*, are colored by obvious religious bias. One story goes that Chastel caught sight of the wolf in the forest, but he was in the middle of praying to the Virgin Mary. He finished his prayers while the wolf thoughtfully waited for him to do so. Then, Chastel shot and killed the wolf. Another has Chastel making his own musket ball of silver for the local priest to bless, a detail that might have been the birth of the legend of the silver bullet needed to kill a werewolf.

In considering the identity of the true beast, most people agree that it was a wolf. Yes, it was a large, vicious, determined, and ferocious wolf, but it was very probably still just a wolf (or wolves). Some insist it was a supernatural being, that no mere animal could have killed so many people in such horrendous ways. The werewolf, of course, is always suggested as an explanation. The killer certainly was particularly cunning and seemed impervious to attacks, even to bullets. Werewolves were said to have glittering, savage eyes, and so did the beast.

Hyenas are put forward as a candidate because of their reddish color. Like a hyena, the beast seemed to be especially heavily muscled in the front with a shorter muzzle and small, pricked ears. There is a striped, modern hyena that fits the description closely. Aristocrats coveted and collected rare animals for their menageries, and animals were known to escape. Hyenas would be unlikely to survive the bitter mountain winters, though.

Jean Chastel had a big, red mastiff dog. At least one researcher thinks his male mastiff mated with a female wolf to produce the beast. That would be physically possible, and would also explain the red/brindle coloration.

My favorite hypothesis, one that best fits transcripts of descriptions taken down at the time the attacks occurred, is a relict population of primordial animals. The area is remote, deeply forested, and inaccessible. If you look at fossils and artists' interpretations of France's indigenous, long-haired hyena, you'll see an odd animal that looks *exactly* like those drawings made in France at the time of the attacks. It was called *Pacyhcrocuta breverostris*. Bones of *Homo erectus* found alongside those of giant hyenas show teeth marks, probable evidence of human predation by the big animal. *Amphicyon ingens* is an equally intriguing contender. With its dog-bear appearance, the shorter, blunt snout, long and luxuriant tail, teeth like a wolf, and limbs that are heavy and strong, this animal could be the poster child for the Beast of Gévaudan. Either one of these ancient beasts would be a perfect explanation for what happened in France from 1764 to 1767. Do I believe either one is likely? Not unless a skeleton that dates from the recent past or a living animal is produced as proof. Unfortunately, the stuffed specimen of the huge wolf with the long tail killed by Antoine and conveyed to Versailles was lost in a fire. It would have been interesting to see what modern science made of the remains of that beast.

What I believe happened in Gévaudan is that a series of particularly savage, heinous attacks by lone wolves, as well as by packs or pairs of wolves, happened at about the same time, causing mass panic. At least some of the attacks were probably perpetrated by wild wolf-dog hybrids, since the two species can and do interbreed. Dogs can produce brindle (striped) coloration. They can be larger than wolves, more stoutly built, and have blunter muzzles.

If you've ever had the hairs on the back of your neck rise when you heard a rustling in the leaves while walking alone through the woods at night, be grateful that you don't live in Gévaudan and that the year is not 1764, for then, the big, bad wolf was real, and he would have shown you no mercy.

CHAPTER THIRTEEN

The Thunderbird

Great wings unfurled, it blocked out the light from the sun like a cloud drifting overhead. High above the earth, so high that you wouldn't have seen her unless you happened to look up at just the right moment, the huge bird effortlessly changed direction with the merest shift of the long, stiff flight feathers at the tips of its wings. Foremost in its mind were the insistent squawks and angry screeches of the fuzzy, downy chicks in the nest. They couldn't fly, couldn't get their own food, and at their fantastic rate of growth, the loud, insatiable demands for sustenance were practically constant. Mice, ground squirrels, and voles were gulped down instantly. Even rabbits were torn to pieces almost before their mother's beak even released them. As soon as the little animals she and her mate brought were consumed, the cacophonous din began again. It was a never-ending cycle. There was no respite; the chicks were insistent, incessant, insatiable.

It was that time of year when the people moved from their summer hunting grounds to the more permanent shelters of their winter quarters. Leaves turned to flame in the forests, a crispness sharpened the air, and everywhere the animals were storing away food for the coming season of want. The people did likewise. There was never enough. Every member of the group who could do so contributed to the effort. Toothless old women smoked meats the hunters had provided for making pemmican. Children barely old enough to walk carefully gathered fruits and berries or nuts from trees or low bushes to mix with the lard and honey that would be added to the cured venison or buffalo meat.

One of the older girls not yet of marriageable age herded her small charges along, carefully putting what they gathered into baskets woven from pine needles or prairie grasses. A chubby toddler ran ahead of the rest crying out with excitement. More berries—and they're ripe! Greedily, he began shoving the juicy black fruit into his mouth. The rest of the children gleefully called out to him and trotted across the meadow to join their leader.

The raptor's eyesight was excellent, the best among all the creatures who flew and far better than any on land. It was able to see a rabbit dive into its burrow from a distance of two miles, a scurrying mouse from hundreds of yards. What it spotted was bigger than a rabbit, big enough to feed her hungry brood who never seemed to get enough to eat. The prey moved slowly, awkwardly. She could sense its vulnerability. As it broke away from the herd, the mother bird seized the chance. Tucking the tremendous wings tight against her body, she pointed her savagely sharp, hooked beak toward the target and plummeted to earth at more than two-hundred miles an hour. The oblivious prey, happily stuffing himself with berries, never stood a chance. His purple-stained mouth was a round O of surprise and pain as the ground disappeared rapidly below his feet. He was still alive as a half-dozen hungry little beaks began to tear him to pieces.

The older girl screamed, a shrill sound to make the blood run cold, pounded toward the great raptor, threw stones at it, but by the time she got to where it had sunk the eight blades of its talons into the tender flesh of the child, it had already lifted itself and its fat prize into the air and was rapidly ascending with powerful beats of its black wings. The rest of the children cried and ran helplessly in circles; had there been another bird in the vicinity they, too, would have been easily picked off.

This kind of prey was something to remember, to hunt for in the future. These creatures were slower by far than rabbits and ground squirrels, had no burrows into which they might escape, and were considerably larger, meatier. They didn't even have fur. The mother bird perched on the edge of the nest, cocked her head to one side, and looked on approvingly as her little brood picked lazily at the gleaming white bones and at last were sated.

Men saw them. They weren't legendary, they were real. The giant predators with wingspans far greater than any of today's birds flew in our skies as recently as 10,000 years ago. Large numbers of animals became extinct between the Pleistocene and Holocene eras, creatures like the mammoths, mastodons, and giant sloths. So did the teratorns with their twelve-foot wingspans and wickedly hooked bills made for ripping and tearing flesh. While condors in California were able to survive the transition by feasting on dead marine animals, those teratorns inhabiting inland areas far from the sea could not successfully compete with hawks and eagles for increasingly scarce food at the end of the last Ice Age, and so they went extinct. The long shadows they cast, though, lingered in the minds of men as stories of those long-ago thunderbirds were passed down through oral tradition among the many tribes of the First Nations.

The original thunderbirds lived long before humans, more than sixty-five million years ago, and their scientific name was Quetzalcoatlus. *Quetzalcoatlus northropi* was named for the Aztec god, Quetzalcoatl, which means "the feathered serpent," and for Jack Northrop, the founder of the Northrop airplane manufacturing corporation. When first discovered, their wingspans were estimated to be as much as sixty feet. Mark Witton, on staff at the University of Portsmouth, United Kingdom, writing in *Geology Today*, January-February 2007, likens their wider wings to those of the condor, while the smaller, narrower wings of the more familiar pteranodon more closely resemble those of an albatross. Later, those early estimates of their wingspans were re-evaluated by Mark Witton himself, along with researchers D.M. Martial, and R.F. Loveridge in "Clipping the Wings of Giant Pterosaurs: Comments on Wingspan Estimations and Diversity," *Acta Geoscientica Sinica*, 2010. They lowered Witton's earlier estimate to a maximum span of forty feet, not sixty.

By comparison, the smallest of the pterosaurs had a wingspan of three-and-a-half feet. According to Wann Langston, Jr., in *Scientific American*, February 1981, pterosaurs couldn't be classified as dinosaurs or birds but were a strange combination of both. "They were flying reptiles that endured for 135 million years. The ones with wingspans of 12 meters (about 40 feet) are thought to have been the largest animals ever to fly."

The scientific bombshell of the existence of Quetzalcoatalus came from Texas and appeared in *Science*, 1975. I well recall the shocking news. "Pterosaur from the latest cretaceous of West Texas: discovery of the largest flying creature," was the title of the article by Douglas A. Lawson, who wrote:

> Three partial skeletons of a large pterosaur have been found in the latest Cretaceous non-marine rock of West Texas. This flying

reptile had thin, elongate, perhaps toothless jaws and a long neck similar to Pterodaustro and Pterodactylus. With an estimated wingspan of 15.5 meters (more than fifty feet), it is undoubtedly the largest flying creature presently known.

In 2010, a study in the journal *PLOS ONE*, a non-profit publication for primary research based in San Francisco, suggested pterosaurs had powerful flight muscles, which they could use to walk on the ground as quadrupeds like vampire bats and vault themselves into the air. A study conducted by Mark Witton and Michael Habib concluded that once airborne, the largest pterosaurs like Quetzalcoatalus could reach speeds of over 67 mph for a few minutes and then glide at cruising speeds of about 56 mph. As they fell from the sky to seize upon prey, their speed approached two-hundred miles an hour.

In an article in the June 2010 issue of *MWP Advanced Engineering* entitled "Engineers make dinosaurs fly," Bassam Ismail wrote: "The gigantic pterosaurs unveiled on London's Southbank as the centrepiece in the Royal Society's 350th anniversary would not be flying overhead without the help of specialist engineers." Mark Witton was involved in making sure the depictions were as scientifically accurate as possible. "The exhibit includes three pterosaurs suspended in flight high overhead and two standing models 20-feet high, making them taller than a double decker bus and tall enough to look a giraffe in the eye. The pterosaurs are the 'wow-factor' of the Royal Society's prestigious ten-day summer science exhibition and were commissioned to help excite interest in science." These extinct giants of the air were formidable and fierce.

In the interview with Ismail, Dr. Witton said:

> No-one has attempted to re-create pterosaurs on such a grand scale before. We've made some of the most accurate models of giant pterosaurs and, so far as I know, we are the only model makers who've taken all modern understandings of their anatomy and footprints into account for our work. Not only are we attempting to show how large these animals were, but we're trying to show how they lived, how they reproduced, fed and fought. From an artistic point of view, I've had to twist my head around complicated aspects of restoration like wing membrane folding, foot placement and making animals that could, theoretically, switch between flying and walking postures.

I was driving into town several years ago across the Butler Bridge that spans the Iowa River a few miles north of Iowa City. Flying just above and in front of me were about six, maybe eight, very large birds in a single-file line. Their heads were quite long and narrow with giant beaks, their wings huge and flapping slowly in unison. It was the strangest sensation, like seeing prehistoric pterosaurs flying out of the past just a short distance ahead of me. I've looked back on that odd experience many times trying to make some sense of it.

Clearly, the birds I saw could not have been what they looked like, which was some kind of prehistoric pteranosaur. Because they were in the air, there was no landmark available from which to estimate scale—to my eye, they appeared absolutely enormous. It's the same problem witnesses encounter when they try to figure out how large a certain shark or whale might be. There's really nothing to compare it with in an endless ocean, just like there was nothing else in the sky that day for me to give the birds some sense of scale. I've come to the conclusion that they must have been pelicans. We get them this far inland; sometimes they number in the thousands up on Lake MacBride or the Coralville wetlands. What I saw *must* have been pelicans, but that experience made me realize how easily someone can mistake the mundane for the mysterious.

In the same way my pelican sighting caused me to do a double-take because they looked so similar to ancient pterosaurs, there *are* some enormous birds still alive in the world. Condors, albeit very few of them these days, still float and drift on air currents high above the arid, coastal mountain ranges of the southwestern U.S. and northern Mexico. Condors are also found in the mountains of South America from Colombia and Venezuela down to the tip of the continent, in Chile and Argentina, although the South American variety is slightly shorter, beak to tail, with a slightly longer wingspan than its California cousins. In the article "Size and Locomotion of Teratorns" by Kenneth Campbell Jr. and Eduardo Tonni, the researchers compare modern condors to their giant, extinct ancestors in the bird family *teratornithidae*. "We estimate that the largest of them, *Argentaves magnificens*, weighed 80 kilograms [about 176 pounds] and had a wingspan of 6–8 meters [20–26 feet]." The extinct teratorns were some two-and-a-half times as large as modern condors. Birds much larger than anything we know today truly did inhabit the skies of an older earth, and people very much like us watched them soar.

In the United Kingdom's *Daily Mail*, July 7, 2014, Victoria Wollaston wrote: The largest bird that ever flew, which had a staggering 24-foot (7 meter) wingspan, was so heavy it would have had to take off like a hang glider. Computer simulations have revealed that the prehistoric condor-like *Pelagornis sandersi* could only have left the ground by taking a running jump downhill into a headwind." Its fossilized remains were unearthed in South Carolina while digging was underway in the construction of an airport. These giants of the sky occurred all over the globe for tens of millions of years but vanished during the Pliocene— just three million years ago. Paleontologists remain uncertain about the cause of their demise. *P. sandersi* breaks the record held by the previously known biggest flyer *Argentavis magnificens*—a condor-like bird that lived in the Andes mountains and the pampas of Argentina six million years ago.

Like my pterosaur/pelican sighting, reports occasionally surface from people certain they have seen abnormally immense, strange birds. Three young boys in Illinois reported being chased across a field by a pair of them. The last boy was said to have been grasped in large talons, lifted a couple of feet off the ground, and finally dropped by one of the birds. The description given by the boys and other witnesses was of birds similar to Andean condors. The two birds were later seen several times by different people in nearby parts of the state. California condors once ranged across much of the country, so it's not impossible that condors could survive outside California, but with legions of avid, active birdwatchers, that possibility is not very likely. Wingspans of the bald eagle, once nearly extinct but now reasonably common in the Midwest, can reach 7.5 feet. Would they attack young boys? Again, not very likely.

Another modern report came from Alaska. On October 19, 2002, several sightings of an enormous bird were reported in an area where four species of eagles, bald, white-tailed, golden, and Steller's, are common. The bird with the largest wingspan living today is the albatross, and Alaska is home to four species of them: Salvin's, Layson's, black-footed, and white-tailed. "A bird the size of a small airplane was recently said to be seen flying over southwest Alaska, puzzling scientists," the *Anchorage Daily*

News reported. The newspaper quoted residents in the villages of Togiak and Manokotak as saying the creature, like something out of the movie *Jurassic Park,* had a wingspan of 14 feet (4.6 meters)—making it the size of a small airplane. "At first I thought it was one of those old-time Otter planes," the paper quoted Moses Coupchiak, 43, a heavy equipment operator from Togiak, as saying. "Instead of continuing toward me, it banked to the left, and that's when I noticed it wasn't a plane."

When scientists weigh in, the claims are put into perspective. "The *Daily News,* the largest daily in Alaska, said scientists had no doubt that people in the region, west of Dillingham, had seen the winged creature, but they were skeptical about its reported size. 'I'm certainly not aware of anything with a 14-foot wingspan that's been alive for the last 100,000 years,' the paper quoted raptor specialist Phil Schemf as saying. Schemf and Rob Macdonald of the U.S. Fish and Wildlife Service said there had been several sightings over the past year and a half of a Steller's eagle, a fish-eating bird that can weigh 20 pounds (10 kg) and have a wingspan of eight feet (2.60 meters), the newspaper reported." The odds of the Jurassic Park bird being a Stellar's eagle or perhaps an albatross seem entirely reasonable.

Stephen Wager wrote an interesting piece called "The Giant Thunderbird Returns" for About.com in February 2016. There were multiple sightings of a giant bird in Pennsylvania going back to 2001. In 2013, this report was made:

> A gigantic bird has been sighted in Pennsylvania. On May 26, 2013, two friends were walking through the woods near Bryn Athyn Castle when they were startled by something extraordinary. "It was extremely loud and I glanced up and saw a huge black bird," Anthony said in his report.
>
> "It was sitting above us and we seemed to startle it. It flew about 100 feet to a nearby branch. Its wingspan was at least ten feet, and judging how far it was it looked to be around four feet tall."
>
> And this was far from the first sighting of such a creature in Pennsylvania.
>
> On the evening of Tuesday, September 25, 2001, a 19 year old claimed to have seen an enormous winged creature flying over Route 119 in South Greensburg. The witness' attention was drawn to the sky by a sound that resembled "flags flapping in a thunderstorm." Looking up, the witness saw what appeared

to be a bird that had a wingspan of an estimated 10–15 feet and a head about three-feet long.

Birds in flight are impressive, large birds even more so. I know from experience that judging their size and species is subjective. Are people seeing eagles and thinking they are condors, seeing albatrosses and thinking they are teratorns? Some reports stretch credibility. Still, it is a fact that before humankind came upon the scene a magnificent giant of a bird did once fly across the skies, blotting out the sun with its tremendous wingspan. It was the feathered serpent, the god of the air. Its name was Quetzalcoatalus. After people spread across the earth, they saw for themselves the nearly-as-impressive birds of prey, the big teratorns, wheeling and turning high in the air above them. Stories of the great birds were told and passed on for generations long after the birds and the people who had seen them with their own eyes were gone. Because their powerful wings beat like the thunder, the people called them thunderbirds.

CHAPTER FOURTEEN

The Kelpie

Janet Mackenzie dawdled along the edge of the loch, now kicking at stones with the tip of her shoe, now pausing to turn her face to the first warm sunshine in weeks. In her hand, she carried a small bouquet of wild bluebells, primroses, and snowdrops. This part of the highlands was battered by fierce winds and driving rain for many months of the year; when it wasn't raining, it was hailing, sleeting, or snowing. She felt her blood singing on this mild spring morning. The new day seemed positively enchanted. Chores at home would still be waiting for her, as she well knew, but for now she wouldn't be rushed.

From up ahead behind a copse of birchwoods, Janet heard a splash loud enough to startle her out of her reverie. What could have made such a sound? If it were a salmon, it would surely be a prize winner. She would remember to tell her brothers about the spot so they might return to fish here. On the well-trodden path she followed, a tall, handsome stranger suddenly appeared from behind the trees.

"Guid mornin', bonnie lass," he greeted her, doffing the flat bonnet worn by men of the highland clans. "Hou ar ye?" A pair of eyes sparkled at her, eyes that were greener than new shoots of the grasses that ringed the loch.

There was nothing untoward about his mild words of greeting, nothing she could put her finger on, but she felt a blush stealing up her neck and across her already pink cheeks. Why was she so flustered? It was a fact that he looked nothing like the village lads with his blue-black curls and flashing green eyes. His clothing was mostly green, too, and not of a type she recognized. Even the clan his kilt represented was one unknown to her. Not only that, it was dripping wet. A few strands of river grass and duckweed were knotted among the dark locks of his hair. Truth be told, there was something timeless about his appearance, some-thing not of this village, maybe not even of this world. Around his neck he wore an intricately wrought silver chain. It drew her eyes. When she raised them to his face, she saw he was smiling in amusement at her obvious discomfort.

"I'll not be talkin' to strangers and have work to do," she told him sharply.

"Ah, but I'm nae a stranger, Janet. I've watched ye walkin' here along the loch many times, ever since ye were a wee lassie."

"Well, I've not seen *ye* spyin' on me!" she retorted hotly and made to step past him on the path.

As she did, a strong arm encircled her waist and drew her to him. Her shawl fell to the ground. She took a sharp intake of breath before her trembling lips were crushed by his. Strong, brown hands held her immobile. Bright green eyes searched hers for a response and found it. She returned his kiss eagerly, although she had never kissed a boy before, and blushed more deeply at her own boldness. This was no boy but a man. Her knees turned to water and her breath quickened. Fiercely, his fingers tangled in her loose hair and they stood gazing at one another as if they'd known each other forever. Time fell away.

All her brothers found of Janet was the shawl she had worn that morning and a scattering of wildflowers. They were lying trod into the mud beside the water. The young men shuddered and crossed themselves.

"Kelpie took her," was the story told and repeated throughout the village, and by nightfall everyone had heard it.

In case you're tempted to consign the legendary kelpie to the rubbish heap of mystical creatures, Benjamin Johnson, writing for *Historic UK, the History and Heritage Accommodation Guide,* writes about a massive modern monument to the kelpie rising up from the ground: "Falkirk in Scotland is home to The Kelpies, the largest equine sculpture in the world. Unveiled in April 2014, these 30-metre [close to 100-feet tall] high horse-head sculptures are situated in Helix Park near the M9 and are a monument to Scotland's horse-powered industrial heritage." The monument is not, however, named Horsepower, but instead is titled The Kelp-ies. The heads are captivating, outsized, and wild. Johnson goes on to describe the animal: "A kelpie is a shape-changing aquatic spirit of Scottish legend. Its name may derive from the Scottish Gaelic words 'cailpeach' or 'colpach,' meaning heifer or colt. Kelpies are said to haunt rivers and streams, usually in the shape of a horse." Some have suggested that the name derives from their strange mane which looks something like kelp but more often like Medusa's writhing snakes.

In *The Heroic Age, A Journal of Early Medieval Northwestern Europe*, June 2005, Craig Cessford, University of Cambridge, published an article titled "Pictish Art and the Sea." In it, he describes stone carvings of a "common symbol known as the Pictish Beast—although originally probably based upon dragonesque brooches—(that) appears to partially be influenced by the shape of dolphins." The Picts were a group of people who lived in northern Scotland between the third and ninth centuries. They carved images of "a sinuous animal with a long snout, spiraled feet and a drooping, typically spiral-ended tail. It is one of the most common symbols in the Pictish repertoire, occurring twenty-nine times..."

These beasts of the water are not only part of Scottish oral tradition but also Irish and Welsh, especially the parts of Ireland and Wales settled by the Celts. Similar stories of water-dwelling demons are shared by Germany, Scandinavia, Central America, and Australia. There are few lakes or rivers of note in northern Britain and in Ireland that don't boast their very own kelpie. These water spirits are far from benign, although anyone who has seen the delightful film *The Water Horse: Legend of the Deep* (2007) might be excused for doubting their malevolent nature. Unlike the film's overgrown, perfectly charming eponymous character, they typically didn't frolic in the bay and allow small boys to hitch a ride.

In the story, young Alex finds a magical egg in the shallows of Loch Ness, which hatches into a ravenous, rapacious little friend who quickly outgrows a barrel and then the family bathtub. The boy and his sister name it Crusoe, after the castaway Robinson Crusoe. Soon, the water horse resembles an extremely over-inflated plesiosaur but with short horns something like those on a giraffe. The film was based on the adorably tame children's book, written by Dick King-Smith (who also wrote *Babe, the Gallant Pig*), and is about as far-removed from the ancient legends as it's possible to get. Crusoe grows rapidly to a prodigious size, taking his rightful place as the next monster of Loch Ness. He is just the kind of phenomenal pet every boy would like to find, with gentle overtones of the early 1960s "Puff, the Magic Dragon" song. Kelpies of legend aren't sweet, though.

Actual kelpies in the old legends and stories handed down over the centuries were a vastly different matter. Come upon a beautiful, black horse, or sometimes a pony, standing at the water's edge and reach out to pet him; the horse was nearly always a stallion. Who could resist? You'd find your fingers stuck fast to the hide of the animal so that you couldn't turn loose. Get up on his back to ride him home and he would plunge beneath the cold, dark waters of the loch or river. There, the beast could devour you at his leisure. The only remaining trace of you found the next day would be your entrails that had washed up along the shore.

One well-known Scottish tale recounts the fate of ten children, all entranced by the magical black horse standing next to the water. Nine of them were unable to dismount. One of the ten, a boy who only touched the stallion with his finger, was prepared. He carried a knife and used it to cut off his own finger, thus escaping the terrible fate of the other nine. In some versions, he loses not one but *all* his fingers, and in still others he must cut off his whole hand to survive.

The original intent of such stories is fairly obvious. The legends are based upon the dual nature of water. Water was life-giving, of course, and necessary for the survival of people, animals, and crops, but it was also dangerous. It could kill children, dragging them down to drown beneath cold, black waves. How to warn them so that the lesson stuck firmly in their minds? The story of the kelpie ought to do it, especially the part about only their guts washing up to mark the spot where they perished. There are also echoes in the old stories of early human sacrifices that were made by the first people to inhabit the desolate Scottish islands in order to appease dangerous water deities.

It was believed that when a kelpie took its victim under the surface, the sound its tail made slapping the water was like thunder. Before a storm broke, ancients believed that the sounds of rumbling thunder and howling winds weren't made by the squall itself but by the kelpies heralding its impending arrival.

Another less common story told in the old times was that the kelpie could assume human form. He sometimes took the part of a rough and raggedy old man lurking in the reeds along the water's edge. As unwary victims would pass by, he would spring forth and again drag them down into the depths where they would be eaten or else crushed to death in his vice-like grip. This kind of kelpie didn't want to befriend children, not in the least. Some legends, perhaps influenced by the spread of Christianity to the region, say that kelpies in human form retain their hooves, sometimes turned backward. This detail links them to the Devil who also had hooves, albeit cloven ones.

Even less often, the kelpie could take the place of a lovely young woman, perched (often nude) on rocks beside the water. You might recognize that she was a kelpie by the water grasses in her tresses, but amorous young men didn't let some wet plants deter them in their pursuit of these enticing maidens. Once she had them in her iron grip, they too would be...well, you get the picture. In prim Victorian times, sexuality was forced to go "underground." The artist's urge to depict and the viewer's desire to appreciate the unadorned female form only met with public approbation if the subject of the painting was a kelpie or a winged fairy or a mermaid. Actual nude females in paintings were strictly off-limits in polite society: fanciful, mythical girls who pranced about with pretty wings or posed alluringly on river rocks in the altogether were considered fair game.

Rarely, the kelpie transformed itself into a handsome young man, a stranger who beguiles some unwary local girl with his handsome visage, masculine physique, and smooth words. He would woo a naïve village maiden, luring her away with him, and then drown her in the water. This variant of the legend was meant to warn impressionable girls—beware of dark, enchanting strangers who, despite their good looks, might harbor nefarious intent.

Scotland's celebrated poet, Robert Burns, wrote of the destructive nature of the kelpies in "Address to the Devil":

> When thowes dissolve the snawy hoord
> An' float the jinglin' icy boord
> Then, water-kelpies haunt the foord
> By your direction
> And 'nighted trav'llers are allur'd
> To their destruction...

When the icy snows upon the waters have melted, the kelpies come forth at the Devil's imperative to prey upon benighted travelers, enticing them by supernatural means to their deaths. It's the old, familiar warning: the bogeyman will get you if you don't watch out! Don't play alongside deep water with strong currents or you may well be dragged down to your death. Don't follow strangers, either, regardless of how pretty or handsome they might be. Some say these Celtic stories were created, told, and retold to warn children and young people of the dangers lurking around every turn in the ancient world, much the same way Sherpas might warn their children not to wander too far from the family hearth for fear of encountering the Yeti.

For all its strength and danger, the kelpie had one Achilles heel—its bridle. Since the usual form it took was that of a horse, the person who managed to control its bridle would be able to subdue and employ the kelpie to his or her advantage. A later, Christian variation was that if a person used a bridle stamped with a cross, he might control the kelpie. The beast had the strength of ten regular horses and stamina beyond imagining, so that incredible amounts of draft work could be accomplished for anyone brave (or foolish) enough to take on a kelpie.

In its human form, if you could manage to relieve a kelpie of his or her necklace, which stood in for the bridle, that would allow you to free yourself from its clutches, although it isn't clear whether young men and women thus ensnared by the kelpie's considerable charms actually *wanted* to be freed or not. One unsubstantiated claim made by Clan MacGregor was that they had an actual kelpie bridle, passed down from generation to generation, when one among them rescued himself from a kelpie on the banks of Loch Slochd by taking the kelpie's bridle, thereby avoiding certain death.

You'll even find kelpies among the magical creatures inhabiting Harry Potter's world. These are the typical, man-eating kind and not the playful come-for-a-jolly-ride kind like Crueso, but they can be subdued by administering a "placement charm" and putting a bridle over the beast's head. The Ministry of Magic puts the kelpie into the category of "beast." They are among beasts ranked in the fourth most dangerous classification, along with Yetis, trolls, and griffins. Dragons, werewolves, and basilisks among others are considered more formidable, but the murderous kelpie is still fairly high on the list of those beasts to fear and avoid. The most famous one inhabits, naturally, Loch Ness.

The terrible nature of the kelpie is described in J.K. Rowling's book, written under the pen name Newt Scamander, *Fantastic Beasts and Where to Find Them* (2001). Know-it-all Professor Gilderoy Lockhart offers Hagrid advice about how to extract kelpies from a well during Harry's second year at Hogwarts, advice Hagrid is wise to ignore. In the boy wizard's sixth year, Harry warns Dumbledore about the possibility of encountering dangerous kelpies in a dark lake that the headmaster must cross in his search for a horcrux.

Whether black horses and ponies with wild manes or handsome men and lovely maids with ebony tresses, it's best to be wary when rambling along the edges of a loch and river where kelpies are said to dwell.

CHAPTER FIFTEEN

The Mermaid

Waves reared up around the ship like a refractory herd of panicking wild horses only to crash back down in a spew of foam. Roiling black clouds raged in the skies above, sending down sheets of driving rain. The sails were already in madly flapping tatters, and it wouldn't be long before the force of the wind would snap the main mast. This was why most sailors never bothered to learn to swim. What was the point? Better a quick death than a lingering one.

The great vessel heaved ponderously onto her side, spilling men like tiny tin soldiers into the angry sea. Howls torn from the throats of the doomed rose up in a long, continuous wail until finally, they were stilled. One man, the valiant first officer, clung listlessly to a bit of floating detritus. His eyes were closed. He was just about to relinquish his hold after asking himself what point was there in prolonging the inevitable when he felt the strange surge of a current beside him. Something broke the surface. Oh, Jesus, no! Please, don't let it be a shark, he prayed. Some deaths were worse than drowning.

His eyes flew open. I must be dead, he thought, as the gentle hands of an angel caressed him, lifting him from the mess of rubble and holding him up. A halo of bright hair floated around her shoulders. Her face was exquisite, her smile a benediction, just like angels in the paintings by the Old Masters. It was the serene countenance from a priceless canvas come to life, a Botticelli, a Van Eyck, a Raphael. This, he belatedly realized, must be heaven. He hadn't expected heaven to be quite so wet. Her lips found his and it was not the kiss of an angel but of a woman, warm flesh and pulsing blood. His eyes widened in surprise. She spoke not a word. There was no need.

The freezing water lashing around him turned languorously warm and was as comforting as any cheerful fire crackling in the hearth at home or a cozy quilt on a chilly night. All at once, drawing air into his lungs seemed utterly superfluous, even ridiculous. It didn't matter and never would again. He sank under the surface of the sea. Rounded pink arms reached out to him, cradling him in their embrace. The fact that his angel had the tail of a fish didn't seem in the least unusual. Why should it? This, after all, was the kingdom of the sea.

Without the merest gesture of protest or fleeting thought of refusal, he dived along with her beneath the rapidly calming surface of the ocean. As the storm had passed over, so had he. He had become one with the sea and all those who dwelt here. Taking the dimpled hand that reached out for his, he kicked once, twice, and felt his legs fuse into a powerful tail that propelled him swiftly through the water. Thoughts of home and family, of the sweet young lady who waited patiently for his return, and of the world of men vanished forever from his mind. Only this enchanting creature swimming beside him mattered. He followed her gladly, willingly, down into the welcoming depths of sapphire blue. She looked back to smile shyly at him, and he realized that his whole life had been leading him to this moment of perfect bliss.

Mermaids sit on rocks near the shore combing long, lovely tresses and singing unbearably poignant songs, songs that can drive a man to distraction. They are beautifully alluring women above the waist but have the tails of fish below the navel. They wear the jewels, pearls, tiaras, and other priceless treasures scavenged from sunken ships. Stories of mermaids have existed in the folk beliefs of many countries for millennia. Sometimes, their natures are benign. They save sailors from drowning, rescuing them selflessly and even falling in love with those they rescued. Other times, they are spiteful and wreak havoc on those who travel on the seas, luring them to their watery graves by singing irresistible songs or by calling up violent storms to destroy sailing vessels. With apologies to merman King Triton, the vast majority of the folk tales, stories, films, and paintings deal with female creatures of the deep.

Most of our Western legends and mermaid traditions are rooted in stories of the Sirens in Greek mythology. Sirens were the daughters of the river god Achelous, sometimes called Akehloios, while their mother may have been Terpsichore, the graceful dancer. Although closely linked to marine environments, mermaids were not considered sea deities. The texts mentioning the Sirens provide different opinions as to their number and their names. Some mention just two or three, while others mention more.

Jason and his argonauts encountered the Sirens, but managed to avoid their dangerous enticements. Orpheus, along on the trip, played the stringed lyre so beautifully that his tune completely drowned out the Sirens' song. In *The Odyssey*, Odysseus was advised by Circe to plug the crew's ears with wax and order them to bind him to the ship's mast. He also ordered the men not to release him, no matter how much he begged, and he apparently begged quite a lot. The crew was able to successfully navigate the waters near the Sirens' island by taking these precautions.

One of the first eyewitness reports of mermaids came from Christopher Columbus on his voyage to the New World, according to Lang Kanai's article, "How Did Manatees Inspire Mermaid Legends?", in the November 25, 2014, issue of *National Geographic*. Columbus wrote this journal entry on his first journey to the Americas. He had caught a glimpse of three "mermaids" off the prow of his ship: "On the previous day [January 8, 1493], when the Admiral went to the Rio del Oro [Haiti], he said he quite distinctly saw three mermaids, which rose well out of the sea; but they are not so beautiful as they are said to be, for their faces had some masculine traits." (Voyages of Columbus, 218). It would have been a very long, lonely voyage indeed to make men so desperate for female companionship that the bristly, mustachioed face of a manatee, the mild cow of the sea, could be mistaken for that of a beautiful woman with the tail of a fish.

As I rode through the J.N. "Ding" Darling Wildlife Refuge on Sarasota Island in Florida one June afternoon, something dark and strange rose from the depths of Tarpon Bay on my right. It was large, you couldn't miss it, and it made a kind of gasping sound as it took a deep breath. The head and shoulders rose up from the water, dripping. The creature hesitated for just a second, and then it dived back down and disappeared with a loud splash of its wide, rounded tail. Clearly, this was something interesting and unusual, at least for me, but I got only a brief glimpse. It didn't stay above the surface long enough for me to examine it closely. Unlike the crew traveling with Columbus, I hadn't been at sea for months living on hardtack washed down with fetid water or warm ale, nor was I suffering from a severe case of scurvy. "Mermaid" came to mind, but only briefly. The creature I saw certainly wasn't attractive in any way. I'm relatively sure that what I glimpsed was a manatee, and I wholeheartedly agree with Columbus' assessment—as far as mermaids go, it was "not so beautiful as they are said to be."

Even William Shakespeare, writing in 1595 or 1596, got into the siren spirit as the Fairy King Oberon, speaking to his servant, Puck, recalls listening to the sweet song of a mermaid, in *A Midsummer Night's Dream*:

> I sat upon a promontory,
> And heard a mermaid on a dolphin's back,
> Uttering such dulcet and harmonious breath,
> That the rude sea grew civil at her song;
> And certain stars shot madly from their spheres
> To hear the sea-maid's music.

Infamous purveyor of humbug P.T. Barnum displayed the Fiji Mermaid (sometimes spelled Feejee) in his Barnum's American Museum in New York in 1842 until the museum burned to the ground. Americans were hoodwinked, certainly, but cheerfully so. Barnum always put on a good show. The torso of a monkey careful sewn to the lower half of a fish was a very long way from the beautiful depictions of mermaids by artists. It was grotesque, shriveled, and hideous, with pendulous breasts and bared teeth. Still, people paid their nickel for the chance to see that wonder of the natural world and weren't disappointed. Harvard University's Peabody Museum of Anthropology and Ethnology claims the original didn't burn in the New York fire and that it still resides in the Peabody collection. Historically, people have wanted to believe in mermaids. They still do.

Film's early fascination with the topic was kindled with *Mermaid* (1911), *Siren of the Sea* (1911), *Neptune's Daughter* (1914), and *Queen of the Sea* (1918), all featuring shapely Australian swimmer and diver Annette Kellerman, the first major actress to appear nude in a Hollywood production. A popular British stage play, *Miranda*, was turned into a film in 1948. It spun a tale of the adventures of a pretty, young, blonde mermaid kept in the bathtub by a married man. It was made into a film starring Glynis Johns and followed up by an even more titillating sequel, *Mad About Men* (1954). Lots of sexual double entendres keep the action lively. The film was set in Cornwall, England, where clergyman Matthew Trewella is supposed to have married a local mermaid.

The Cornwall Guide says:

> The small village of Zennor huddles around the medieval church between the West Cornwall moors and North Cornish coast not far from St Ives. In that church carved on the end of one of the wooden pughs [pews] is a strange figure of a mermaid. Depicted with long flowing hair, holding a mirror in one hand and a comb in the other, is the Mermaid of Zennor.

She is reputed to be one of the daughters of Llyr, king of the ocean, and was named Morveen. It's no wonder many folk tales of mermaids have arisen in the area with that as inspiration.

Hollywood's *Mr. Peabody's Mermaid* (1948) has married man William Powell irresistibly drawn to gorgeous (although somewhat wide-eyed and dim-witted) brunette Anne Blythe who sings her little heart out. In *Splash* (1984), Tom Hanks finds himself enthralled by Daryl Hannah, who hangs out on occasion in the bathtub and eats lobsters in a manner that would appall Emily Post. Disney, never a company to miss an angle, came out in 1989 with an extremely modified (no cutting out of tongues, no turning of heroine into sea foam at the end) version of the Hans Christian Andersen story *The Little Mermaid*, and the mild, 'tween offering *The Thirteenth Year* a decade later.

Australia jumped back into the deep end of the ocean with *Aquamarine* (2006) and a couple of teen TV series with mermaids as main characters, *H20: Just Add Water* (2006) and *Mako: Island of Secrets* (2013). I admit I've enjoyed every one of these films and television shows multiple times, except those earliest Kellerman films, which I have yet to watch and, like many people, I sincerely wish mermaids were real.

There's just something about a girl in a neoprene tail. There were mermaids at Disneyland on the Submarine Voyage from 1965 through 1967 until one too many sailors on shore leave jumped into the lagoon to heed their seductive siren call. At least that's *one* version of the story. Another more prosaic explanation is that the chlorine in the water and diesel fumes from the subs created unsafe working conditions. Sara Boboltz, the entertainment editor for *Huffington Post*, created a fun feature on the Disney mermaids posted on July 25, 2015, including old videos. She wrote, in an article titled "A Brief History of That Time Disneyland Employed Live Mermaids," that after photographs featuring "Disneyland's performing mermaids surfaced on the Internet a few years ago, one former sea nymph recalled a peculiar question a park monorail conductor once fielded from an elderly visitor. 'Are they real?' she asked of the mermaids. The conductor responded in the affirmative, to which the visitor mused, 'I wonder where they found them. Probably the Sargasso Sea.'" Maybe it was said tongue-in-cheek, maybe not. Remember, people *want* to believe in mermaids. They always have.

More recently, "The Body Found" was a fantastically successful documentary-style program produced by Animal Planet that first aired on May 26, 2013, showing viewers that not only are mermaids real, but there is an awful lot of irrefutable evidence to prove it. The show was followed by a second installment, "Mermaids: The New Evidence." At the very end of the first program, there was a brief, deliberately insignificant disclaimer saying something along the lines of, "Oops, just kidding, folks—this has all been a big load of hooey!" The scientists were played by actors. The evidence was manufactured. As someone who watched the first airing of the program, completely unaware that the program was meant to deceive, I can attest that not only was it extremely convincing, it was also scary as hell.

The mermaids that appear living in the Great Lake surrounding Hogwarts in Harry Potter's fourth book and film, *Harry Potter and the Goblet of Fire*, are also viciously dangerous. Remember the clue? "Come seek us where our voices sound, we cannot sing above the ground." They certainly aren't beautiful, with their gray-green skin, yellow eyes, and teeth like a piranha's. Those pointy spears they carry mean business, too.

Marine biologist David Schiffman wrote an article titled "No, Mermaids Do Not Exist" shortly after the Animal Planet documentaries aired. He says:

> I can tell you unequivocally that despite millennia of humans exploring the ocean, no credible evidence of the existence of mermaids has ever been found. Sure, new species are discovered all the time, but while a new species of bird or insect is fascinating, it doesn't mean "anything is possible."

Schiffman quotes deep-sea ecologist Andrew David Thaler:

"Look, the ocean is a vast, unexplored frontier. The deep sea is Earth's last great wilderness. When we do venture into the abyss, we find creatures more diverse and incredible than our relatively limited imaginations can conceive. Don't insult that wonder with something as utterly mundane as "human with fish tail.""

Mundane, maybe, but gaze for a moment at the haunting loveliness portrayed so believably in "A Mermaid," an oil painting by Royal Academy member John William Waterhouse done between the years 1895 and 1905. Part of a private collection for many years, it resurfaced in the 1970s and is once more part of the academy's collection. The siren Waterhouse created is exactly the way we hope a mermaid might appear—red-gold flowing hair, an abalone shell nearby spilling over with pearls, skin like porcelain, and a pensive expression of such aching, otherworldly beauty that it would surely melt the hardest heart. What man could resist her call? More to the point, what man would want to?

CHAPTER SIXTEEN

The Mahamba, the Neguma-monene, and the Mokele-mbembe

Two little sisters stood by the water's edge, shoving each other playfully and teasing. Their necks were simply adorned with a few corded necklaces. That was all they wore. It was hot, and modesty was not a concept in that part of the world in which they lived. They were only a year apart in age, but the older of them felt she had earned the right to be in charge. They carried earthenware jars to fill with water and felt very important with the weight of this responsibility. Despite the sun dropping toward the horizon, the jungle was still steamy, and their damp skin glistened from exertion. Their faces were running with rivers of sweat. At this rate, they'd be lucky to get back to the village before dinner time.

The smooth surface of the lake was placid. It was the time of day when thirsty animals would begin to make their way down to its banks to drink. The golden wolf, African bush elephant, cheetah, giraffe, chimpanzee, zebra, jackal, hyena, Congo lion, eland, bonobo, mountain gorilla, okapi (the animal once assumed to be mythical), and a whole host of other wildlife inhabited this remote region along with the little sisters and their people, who were also very little.

The water was also home to crocodiles, hippos, and even dugongs, the African variety of manatees, who swim up the rivers from the sea as far as they are able to pass. Giant pythons, Gaboon vipers with two-inch-long fangs, and the most lethal snakes of them all, black mambas, draped themselves quietly over branches in the trees. This lush, dangerous land was home to the girls, but they shared it with a rich variety of other creatures, the docile as well as the deadly. They had been taught from babyhood to afford the other dwellers in their environment a healthy respect.

A little herd of the sweet-natured, gentle upemba lechwe came stepping daintily from the dense undergrowth. The boldest among them shoved its way to the front of the group with the judicious application of beautifully curved, slender horns. The deer weren't very tall and offered no threat, so the two children relaxed and watched.

From the depths of the lake, a tremendous black shape propelled itself toward the largest of the pretty beasts. The girls realized instantly that the animal was doomed. Jaws were already agape as the beast hurled itself out of the water. Ugly yellow spikes of teeth, each one as long as the span of a man's hand, closed around the delicate head of the frantic herbivore as it was bent to drink. Nearly as quickly as it rose from the depths, the predator withdrew back into the water with its prey.

Calmly, even lazily, it swam away from the bank propelled by the unhurried, swaying movements of a long, powerful tail. Over and over it rolled until the sharp hooves and horns ceased to thrash. The surface of the lake was as quiet as it had been a few moments before. The rest of the upemba lechwe drank thirstily, crowding each other for room at the water's edge. It was as though nothing at all had just happened.

The girls filled their vessels with water and turned away, glad that the herd came down to drink when it had. Otherwise, one of them would have surely been returning to the village alone.

The Congo has spawned a variety of legends telling of prehistoric beasts, stubborn holdovers from the age of the dinosaurs. Like oblivious party guests who simply refuse to leave as the festivities wind down, these creatures may have overstayed their welcome, but they have provided in exchange colorful stories for generations of locals who live in this remote region. There are reports of sightings of the mahamba, a giant crocodilian-like reptile, a fresh-water version of the prehistoric mosaasaur, an animal presumed to be extinct from the end of the Cretaceous period. One legendary beast, the neguma-monene, resembles a variety of stegosaurus. Another, mokele-mbembe, looks like a small brontosaurus. In fact, if eyewitness reports can be believed, and unfortunately they can't, nowhere else on earth can lay claim to quite as many fantastic creatures as the remotest regions of the African Congo—unless it is perhaps the swamplands of the southeastern United States.

With the number of monsters living in the swampy Lake Likouala region of the Republic of the Congo, it would seem that adventurous tourists or explorers hoping to catch a glimpse of one would be positively tripping over likely candidates. This is not the case. Film taken of the creatures is grainy and not sharp enough to decipher much at all, other than a basic dark shape. Occasionally, promising footage was spoiled when a lens cap was inadvertently not removed. Well, that *can* happen. Footprints have been found, but footprints are not enough evidence and are easily subject to hoax. Remember Nessie and the hippo-foot umbrella stand? Bigfoot? Yeti?

Many of the oldest of the legends, and some have been recorded as far back as the sixteenth century, are traced to the pygmies who inhabit the region, members of the Aka, Efé, Twa, and Mbuti tribes. There are twelve distinct groups of them. Adult males in these groups are nearly all under five feet tall. Pygmy is a word from the Greek *pygmaios,* now sometimes considered a derogatory term. No single word has yet emerged, however, to cover the various groups, although sometimes Bayaka is used.

As to the mahamba, the simplest explanation may be the correct one: if it looks like a duck, swims like a duck, and quacks like a duck, then it's probably a duck. This form of abductive reasoning follows logical inference to its most likely conclusion. The largest saltwater crocodiles, *Crocodylus porosus,* may reach twenty-three feet in length, larger than the biggest of great white sharks. They are the greatest terrestrial and riparian hypercarnivorous apex predators on earth. They are aggressive. They can swim out to sea for hundreds of miles and occasionally cruise along warm water currents for up to a thousand miles. These real-life "monsters" can be found in northern Australia, India, and Southeast Asia.

The mahamba, a crocodile-looking creature, is to the modern saltwater croc what megalodon is to the great white shark. Reports by Bobangi aboriginals claim lengths of fifty feet for the swamp-dwelling mahamba. They compare it with the nkoli, their word for crocodile. It is fairly easy to imagine the scenario where an enormous, aggressive, territorial male "salty," as the saltwater crocs are familiarly called, would scare the very daylights out of anyone paddling along in a slim canoe on Lake Likouala. Salties are known to attack small water craft. Following "the one that got away" principle, stories of the gigantic beast would grow in the retelling of the tale. It would have certainly seemed formidable enough to merit being considered a monster to anyone who encountered it. Was it a genuine prehistoric beast? Crocodiles have been around for at least four million years, based on fossil records. They *look* prehistoric. Until a mahamba is caught, however, it's impossible to know with any degree of certainty just what the lake monster is. Still, if it *looks* like a croc...

People who believe the "relict population" theory speculate that the creature is a long-lost relation of ocean-going mosasaurs. Somehow, they suggest, the saltwater mosasaurs have been stranded in fresh water and have adapted. Since bullsharks, tarpon, and sawfish are routinely found in freshwater Lake Nicaragua, there is some actual precedent for this idea.

The neguma-monene resembles a stegosaurus. To date, the two villages of Bounila and Ebolo are the epicenter of sightings. The creature has been described as having planks coming up from its back, planks covered in green algae. No physical evidence of the beast exists. It is at least semi-aquatic, looks like a stegosaurus (more specifically a Kentrasaur), and eats plants. Crocodiles have rigid scutles, specialized scales covered in strong beta-keratin, that are especially prominent along both sides of its tail. Those might look like planks to a terrified observer paddling madly in the opposite direction.

Many Kentrasaur fossils were found in Tanzania in East Africa, and the animal was given its name as recently as 1915. Tanzania shares its western border with the Republic of the Congo, divided from each other by Lake Tanganyika. The proximate geography would make the existence of a Kentrasaur in the Congo at some point in time seem at least possible. A large number of its fossils have been found since the beginning of the twentieth century. If fossils of such a spiked beast were ever seen by locals, recently or in the distant past, then linking those fossilized remains with living creatures people had seen in the swampland would explain a possible source of the confusion.

Neguma-monene means large python in the Lingala language. Large python does not seem a very accurate description for a modern-day Kentrasaur. There were two recorded sightings in 1961 and 1971 by Joseph Ellis. It did not seem much like a snake at all and was more lizard-like. He compared the length of its tail to the boat in which he rode as being about thirty-feet long. He didn't see the head because it was submerged. The subject was not one the locals wished to discuss, which seems rather disingenuous, given the number of legends originating from this little corner of the world. Both sightings were made along the Dongu-Mataba, a tributary of the Ubangi River.

The last of the three famous creatures of the Congo is the smallish brontosaurus called mokele-mbembe. Roy P. Mackal is most closely associated with prehistoric monsters of the African Congo region. He served as a marine in World War II, earned a Ph.D from the University of Chicago, and went on to teach there as a professor. His main work was done in biochemistry and virology. For several decades, Mackal performed influential research in the study of viruses, studying bacteriophages and the lysogenic cycle. He later served as a professor of zoology at the same university, retiring from academic life in 1990. He became interested, like so many of us did, in reports of the Loch Ness monster and subsequently in the mysterious creatures of the Congo. He was one of the founders of the now defunct International Society for Crytozoology. Professor Mackal died in 2013.

Along with University of Arizona ecologist Richard Greenwell and Congolese biologist Marcellin Agnagna, Mackal made two expeditions between 1980 and 1981 in an attempt to find and photograph the elusive mokele-mbembe. Its name, literally translated, means one who stops the flow of rivers. Mackal did not see one, but later detailed his research and compiled interviews of those who said they had in his 1987 book, *A Living Dinosaur? In Search of Mokele-Mbembe*. By all accounts, Roy Mackal seems to have been quite a decent guy, well educated, highly motivated, and acting with good intentions. He didn't make claims he couldn't prove, and he couldn't prove the existence of the little brontosaurus. Neither could the other two dozen expeditions launched for that same purpose.

What we are left with are anecdotes, all of them unscientific, most of which have been meticulously recorded, at least in recent years. It isn't completely beyond belief to imagine that strange beasts may still lurk undetected in the Congo River basin. No teeth, no bones, no carcass, and no decent filmed records, however, don't make their existence especially likely. Are fifty-foot crocodiles, still-living Kentrasaurs, and very manageably-sized, river-dwelling sauropods playing a coy game of hide-and-seek in the Republic of the Congo with legions of determined explorers? Isn't their existence at least a remote possibility? Before you answer, recall that until the middle of the nineteenth century, gorillas were considered to be merely legends, too. When they were discovered there was no word for the impressive hominid in English—or in any other modern language.

CHAPTER SEVENTEEN

The Nandi Bear

She was the last of her kind still roaming the mountains of Africa. In her youth, she had found a mate and then had young to care for. Those days were gone. Now, she had lived many seasons without ever coming across another like herself. Predators were non-existent, for who could challenge her? She was big, strong, and not something to be trifled with. Her nature was mild, but if provoked, she was a formidable opponent. Weapons that could bring her down hadn't been invented until quite recently, but, of course, she had no way of knowing this. She had no problem finding enough to eat. Grubs, berries, fruit, and the occasional small animal made up most of her diet. Those things were plentiful in the region and kept her fat. Companionship of another creature, however, was a pleasure only dimly remembered.

"Look!" the hunter said under his breath to the helpers who followed behind him. "Scat!" Yes, he had found the unmistakable, fresh spoor left by his quarry. It was unlike anything else, a tell-tale sign that he was hot on the trail. "Ah, and here's a track in the mud. We've got it!"

The temperature never dropped too far in this region, but the nights were growing cooler. She lumbered toward the small, protected cave in the crevice on a hill where she slept. The sun was setting. It cast a rosy, golden glow over the countryside throwing everything into soft focus. Her dark, bulky shape was clearly visible against the pale, dried grass on the hill. She chuffed softly with contentment as she approached the comfort of her den. In a few minutes, she would fall into a dreamless sleep, curled into a big, furry ball, and she would snore softly.

One of the bearers pointed toward the hillside. "Boss," he hissed.

"I see it!" This is what he had come for, the trophy of a lifetime.

He lifted the heavy rifle onto his shoulder and took aim. There was absolutely no danger to himself or to any of the local men he hired to accompany him on the hunt. The animal, even if badly wounded and angry, was too far away to pose any serious threat. No, he would be perfectly safe no matter what happened. He smiled smugly in satisfaction.

BAM!

The loud shot rang out, and the reverberation could be heard for miles. The big rifle recoiled powerfully into his shoulder nearly knocking him to the ground. Birds gone to roost for the night squawked in hoarse alarm and rose in a cloud of flapping black wings into the twilight sky.

"Got it!" he exulted, proud of his kill. "Did you see that?" he asked, grinning widely and looking from face to face as if the others might have somehow managed to miss it. Had he known at that moment that he had just killed the very last of her kind in all the world, his swelling pride would surely have been even greater.

Africa is home to yet another mysterious creature named the Nandi bear because of its range in Kenya, East Africa, where the Nandi people live. The majority of reports come from the western part of that county. The usual suspects for the Nandi bear are identified as a bear, hyena, or perhaps some sort of relict population of extinct animal such as a chalicothere, this last hypothesis being suggested by no less an authority than famed paleontologist Louis Leakey himself. The Nandi bear is a nasty-tempered, viciously opportunistic carnivore you wouldn't care to meet alone in a dark alley. With long, heavily-muscled front legs, a blunt muzzle bristling with sharp teeth, and a sloping back that ends in shorter hind legs and a stumpy tail, this is one nasty animal. It has a brownish-colored coat and stands nearly four feet tall at the shoulder. Not only that, the locals say that when it takes a human victim, it only feasts on the brain and leaves the rest. The elusive Nandi bear is about as far from a cuddly Teddy bear as you can get.

While Nandi bear is its most common name, the animal is also called the Duba (from dubbha, a combination of bear/hyena in Arabic), Chemosit, Kerit, or Ngoloko, and there are various other names, too, depending on the traditions of a particular local region where the creature is believed to live. As you've no doubt noticed, many people assume any odd or unusual beasts must be a relict population of extinct creatures, and the Nandi bear is no exception.

Some hypothesize that sightings of the Nandi bear as reported by witnesses indicate that it belongs to a group of prehistoric hyenas, namely *Pachycrocuta brevirostris*, known more familiarly as the giant hyena. That animal came on the scene some three million years ago in the late Pliocene and disappeared during the mid-Pliocene some four hundred thousand years ago. It was a big predator similar in size to a female lion. The overdeveloped front legs and chest fit the descriptions of the Nandi bear. Fossil remains show that *Pachycrocuta* stood about three-and-a-half feet tall at the shoulder and probably weighed up to 250 pounds. Remains of the animal in China's Zhoukoudian Cave show remains of *Pachycrocuta* found alongside those of *Homo erectus*. Marks on the bones of early man were at first supposed by researchers to indicate cannibalism, but they now are believed to show predation by the prehistoric giant hyenas. The group of *Pachycrocuta* was probably out-hunted by the quicker, smaller spotted hyena from which modern hyenas descend.

If the animal *is* actually a bear and not just a bear-like animal, did bears ever live in Africa? The answer is yes. Brown bears called Atlas bears, *Ursus arctos crowtheri*, did exist on the African continent in modern times, but it was not a native population. Bears are good swimmers, especially polar bears who are genetic relatives to the Atlas bears, and bears could have swum from the Iberian Penninsula to North Africa fairly easily. Also, Rome brought in Cantabrian brown bears, *Ursus arctos arctos*, living in the Cantabrian Mountains in the north of Spain. The bears were used for bloody spectacles and contests in the Colosseum and lesser arenas for public entertainment. Bears fought each other, lions, tigers, bulls, and were even used as a means to execute criminals. They were treated cruelly and starved before such events. Rome's reach extended into the northern part of Africa, and small breeding populations of brown bears became established in the region. They were believed to have been hunted to extinction by the 1800s, with 1870 the date of the last recorded death of the wild Atlas bear, in the mountains of northern Morocco. Once rifles were powerful enough to shoot and kill bears were invented in the 1850s, the fate of the Atlas bear was sealed.

In *Bruin, the Great Bear Hunt* by Mayne Reid, written in 1864 and published in 1874, the death of one of these animals is recorded a few years before the last of them was killed:

> One of them that was killed near Tetuan, about twenty-five miles from the Atlas Mountains, was a female, and less in size than the American black bear. It was black also, or rather brownish-black, with no white about the muzzle, but under the belly its fur was of a reddish-orange. The hair was shaggy, and about four or five inches long, while the snout, toes, and claws were all shorter than those of the American black bear, while the body was of thicker and stouter make.

In fact, this female was one of the very last remaining individuals of her breed, although the author probably didn't realize it at the time he wrote about the hunt. The careful description provided by Matthew Reid gives us an excellent comparison with reports of the Nandi bear. Stout, short-snouted, brownish fur with reddish underbelly are compatible with eyewitness reports. Early reports of the Nandi bear were possibly Atlas bears. Their nature was relatively gentle and shy, and they ate roots, berries, acorns, and fruits, for the most part. That part does *not* fit the vicious reputation of the Nandi bear.

Louis and Mary Leakey are most famous for their work in establishing the emergence of early man in Africa, due to discoveries they made at the Olduvai Gorge in Tanzania. Louis Leakey also speculated that the chalicothere might be the inspiration for stories of the Nandi bear. He wrote an article for the *London Illustrated News*, November 2, 1935, titled "Does the chalicothere— contemporary of the okapi—still survive?" The chalicotheres lived from the middle Eocene to early Pleistocene eras, disappearing from the planet about 780,000 years ago. Why would the highly respected Dr. Leakey think they might still be living in Africa? His belief was based on the fact that the physical description of the chalicothere was similar to that of the Nandi bear. They had long, muscular front legs and smaller hind legs, but beyond that, there isn't much overlap. Chalicotheres were browsers, herbivores, and resembled an odd combination of horse, tapir, and rhinoceros, all of which were its closest relatives.

Leakey speculates that the chalicotheres survived in East Africa until approximately 12,000 years ago and possibly longer than that:

These animals were horselike mammals with retractable claws instead of hooves, long necks, rearward-sloping backs, and long front limbs that were much longer than the hind legs. They browsed on tree leaves and probably knuckle-walked like a gorilla, with their claws curled inward. Evidence that a chalicothere may have survived into modern times comes from a Saka (Scythian) tomb in Siberia (400–500 BC), where two gold belt plaques were found that showed a horse-like animal with clawed feet.

After looking at the artifacts discovered in Scythian burial mounds, many displayed in the Hermitage Museum in St. Petersburg, Russia, there is no doubt that the Scythians knew the difference between modern horses and other animals. There are horses, deer, a panther, a tiger, and many other animals accurately depicted in gold, and then there is also an odd, horselike animal with claws. This is pretty slim evidence, but it is all there is at this point. The okapi wasn't recognized as an actual, as opposed to mythical, animal until 1901 when the skin and skull of one were taken to London as proof of its existence.

Indigenous people have seen the Nandi bear for hundreds of years, but it was first reported by colonists in the early 1900s. In the *Journal of East Africa and Uganda Natural History Society*, Vol. 4, 1912, Geoffrey Williams wrote an article called "An Unknown Animal on the Uasingishu" about his personal experiences. He was a member of the Nandi Expedition and published this observation:

> I was traveling with a cousin on the Uasingishu just after the Nandi expedition, and, of course, long before there was any settlement up there. We had been camped...near the Mataye and were marching towards the Sirgoit Rock when we saw the beast... I saw a large animal sitting up on its haunches no more than 30 yards away... I should say it must have been nearly 5 feet high... It dropped forward and shambled away towards the Sirgoit with what my cousin always describes as a sort of sideways canter... I snatched my rifle and took a snapshot at it as it was disappearing among the rocks, and, though I missed it, it stopped and turned its head round to look at us... In size it was, should I say, larger than the bear that lives in the pit at the "Zoo" and it was quite as heavily built. The forequarters were very thickly furred, as were all four legs, but the hindquarters were comparatively speaking smooth or bare...the head was long and pointed and exactly like that of a bear... I have not a very clear recollection of the ears

beyond the fact that they were small, and the tail, if any, was very small and practically unnoticeable. The color was dark...

Railroad workers in the same area also reported seeing the unusual animal and one of them got a fairly good look at it. In the journal cited above, Vol. 6, 1913, the engineer G.W. Hicks contributed an article, "Notes on the Unknown Beast seen on the Magadi Railroad." He reported, on March 8, 1913, that he saw something about the size of a lion with strongly muscled forequarters, small ears, high withers (shoulders), and a blunt muzzle. It reminded him of the reports given by railroad workers as well as civilians living in the area.

Other animals suggested as contenders for the Nandi bear are the giant forest hog, wild dog, aardvark, spotted hyena, badger, or perhaps even some kind of prehistoric baboon or hitherto unknown sort of hominid.

Until a Nandi bear is actually captured, there will continue to be debate not only about its origins but over its very existence. Of course it's also possible that *any* strange, briefly glimpsed, and unidentified animal in the region will be classified as a Nandi bear for want of a better candidate. I tend to think that those early reports were of real bears, probably the Atlas bear, while more modern sightings could have been just about any or all of the above candidates—except for the chalicothere, with apologies to Dr. Lewis Leakey.

CHAPTER EIGHTEEN

The Tasmanian Tiger and the Queensland Tiger

Every state and territory in Australia has its own unique set of hunting regulations, and he had every one of them committed to memory. This year, he was making his living in the northern territory of Queensland where no recreational hunting at all was permitted, with one key exception: "Hunting is limited to feral animals on private property with landowners' permission to hunt on the property. Only a current firearms licence is required to hunt on private property. There is no hunting permit or fee applicable." Only animals classified as "pests" could be taken, and that's how he earned his pay. He was an exterminator, a good one. A farmer complained to him, they agreed on a payment, and he'd rid the land of whatever animal was causing problems. Donkey, deer, dingo—it was all the same to him. Feral pig, fox, hare, or wild goat—they all came under the designation of "pest," and he was happy to help...for a fee, that is.

He crept through the shady underbrush, barely making a sound. Nothing he hunted was very large. A 12 gauge was plenty big enough for all his needs, loaded with 00SG buck. He chose a sleek and lightweight synthetic Benelli M2 with inertia action. If he aimed, he rarely missed. The powerful weapon with a red dot sight he had mounted was slung casually over a shoulder. He barely noticed, it had become so much a part of him. Hunting almost since he could walk, he was able to step on a dry twig without making it crack, able to move as stealthily and quietly as the animals he tracked. Had to. If they knew you were coming, they wouldn't hang around waiting to be shot, and if he didn't shoot them, he didn't get paid. Mosquitos hummed in his ears, but he paid them no heed. A sheep farmer had called him yesterday about a dingo hanging around picking off young stock. The price of lambs these days had made it worth his while to hire a professional. Glad to oblige.

About fifty yards ahead on his left, he caught a flash of dun-colored fur against the deep green foliage, followed by the lash of a long tail. *There you are,* he thought with satisfaction. He lifted the shotgun and took aim. No sweat. He'd be able to grab his pay for the dingo and be on the way to picking off those feral pigs a couple of farms over by this afternoon.

The animal in his sights stopped and turned to look around warily, suddenly on high alert. Some sixth sense seemed to kick in warning it that danger lurked in the forest, and danger was something an apex predator seldom had to worry about. It yawned nervously, just like modern dogs will do when they become anxious. The jaw of the creature nearly came unhinged. A mouthful of sharp teeth showed clearly as the animal opened an unnaturally wide mouth. It took a few hesitant steps and paused. The hunter saw wide, dark stripes across its back. It had that characteristic waddle of a marsupial, the shortened, cramped-looking rear legs. This was no dingo. He was looking at the living history of Australia's past. He felt gooseflesh rise on his forearms. Bringing back the dead animal would prove its existence once and for all. It would also propel him to instant fame—like the dentist from America who killed Cecil the Lion.

Noiselessly, he shrugged the lethal weapon over his shoulder and headed back the way he had come. He wouldn't report the sighting. The animal deserved a fighting chance. Back at the farmer's residence, he heaved the body of a recently shot dingo from the back of his Ute onto the dusty driveway, holding out his hand to receive payment. No sense being an idiot about it.

Australia is home to a considerable number of unusual species of wildlife, including marsupials. From the large supercontinent known as Pangaea, the landmass that included Australia, Antarctica, South America, and Africa separated because of the movement of tectonic plates some two-hundred million years ago. From that secondary landmass, Australia broke free about one-hundred-and-fifty million years ago, and its wildlife developed largely independent of influences from neighboring lands. Among the best known of its marsupials are the kangaroo, koala bear, wombat, Tasmanian devil, and opossum. Many people are also familiar with the unfortunate history of the thylacine, better known as the Tasmanian tiger or Tasmanian wolf.

The range of the now-believed-to-be-extinct thylacine was limited to Australia, Tasmania, and New Guinea. It was largely nocturnal and shy, an apex predator with many characteristics of modern dogs. Its coat was fawn, yellow, or brown with thirteen to twenty-one darker stripes on its back. Thylacines could be up to fifty-one inches long with tails up to twenty-six inches. They stood about twenty-four inches at the shoulder and weighed up to seventy pounds. Although their jaws were relatively weak, they were able to open them up to eighty degrees.

Take a few minutes to look at some of the existing films made of captive thylacines. The films are both fascinating and heartbreaking. The last footage was taken in 1933 by David Howells Fleay, a naturalist who pioneered the breeding of endangered species at Hobart's Beaumaris Zoo in Tasmania. It shows what is widely considered to be the last living representative of the species. My parents were teenagers when the last of the thylacines died, making its demise relatively recent. It is little wonder people want to believe that perhaps a few breeding pairs may yet survive in the farthest reaches of Tasmania's backcountry. To think humans hunted such a mild, fascinating creature off the face of the earth is a terrible indictment on the excesses of our domination over the rest of nature's creatures. Search "last living Tasmania tiger" to see poor Benjamin in his small, barren cage pacing back and forth and opening wide his jaws. According to an article by Amy Ziniak, writing for the *Australian Daily Mail* on September 6, 2013, Benjamin was left out in the cold one night, locked out of his sleeping quarters, and he froze to death as a result of this act of neglect. He died on September 7, 1936. As Mark Twain so perfectly summed us up, "Man is the only animal that blushes. Or needs to."

The Parks and Wildlife Service in Tasmania published this timeline of the thyacine's demise:

- 1830: Van Diemens Land Co. introduced a thylacine bounties.
- 1888: Tasmanian Parliament placed £1 bounty per thylacine head.
- 1909: Government bounty scheme ends. 2184 bounties paid.
- 1910: Thylacines rare—sought by zoos around the world.

- 1926: London Zoo bought its last thylacine for £150.
- 1933: Last thylacine captured, Florentine Valley, sold Hobart Zoo.
- 1936: World's last captive thylacine died in Hobart Zoo, July 9.
- 1936 Tasmanian tiger added to the list of protected wildlife.
- 1986: Thylacine declared extinct by international standards.

Naturalist John Gould correctly predicted the end of the thylacines in 1863:

> When the comparatively small island of Tasmania becomes more densely populated, and its primitive forests are intersected with roads from the eastern to the western coast, the numbers of this singular animal will speedily diminish, extermination will have its full sway, and it will then, like the wolf in England and Scotland, be recorded as an animal of the past.

The Tasmanian Parks and Wildlife Service weighed in with this opinion:

> Every effort was made, by snaring, trapping, poisoning and shooting, to fulfill his prophecy. Bounty records indicate that a sudden decline in thylacine numbers occurred early in the 20th century. Hunting and habitat destruction leading to population fragmentation are believed to have been the main causes of extinction. The remnant population was further weakened by a distemper-like disease.

Still, sightings continue to be reported. An adventure film staring Willem Dafoe and Sam Neill called *The Hunter* (2011) is a fictionalized depiction of such sightings and how they might conceivably encourage hunters hoping to kill "the last one." It pits those who hope against hope that the creatures may yet survive against those who would ensure its complete destruction.

Many sightings are considered "good," yet none have been conclusive. There have been several searches for the animal, but none that have been successful, including those using motion-activated cameras. While the thylacine may be the only mammal to become extinct in Tasmania since Europeans arrived, fifty percent of Australia's mammals became extinct during that same period.

The Queensland tiger was first documented in the records in 1871, but there were numerous sightings in Australia from indigenous people before that of an animal they called the yarri.

It is reported to be about the size of a big dog, but unlike the mild Tasmanian tiger, the Queensland tiger is said to possess a fierce temperament and very prominent, sharp fangs. The animal is striped, fast, and agile. The name Queensland comes from the northeast region of Australia where most of the sightings have occurred. *Thylacoleo carnifax* means "meat-cutting pouched lion," an extinct, carnivorous marsupial; the last of them roamed Australia some 30,000–45,000 years ago. They would likely have overlapped with the early human settlers. Some scientists think as men hunted the large animals on the continent to extinction, the main sources of food for *Thylacoleo* would have disappeared. In addition, people burned the ecosystem necessary to support the big cats. As with many strange animals, some say that perhaps a relict population of these ancient predators could have survived and continued to live in the most remote regions of the land.

The Centre for Fortean Zoology Australia reports:

> Not everyone is prepared to dismiss the Tweed legend of a large flesh-eating marsupial lion stalking the dense hinterland near Tumbulgum, with a handful of locals convinced the beast is real—according to the *Gold Coast Bulletin*. For years an urban myth about this ancient creature has circulated on the Tweed. But experts reject the theory, saying *Thylacoleo carnifex* (murderous lion) was too long extinct. Fossils indicate the marsupial lion was the largest meat-eating mammal known to have ever existed in Australia. Recent remains of it have yet to be found.
>
> Tweed Historical Society member Brian Boyd said he heard the stories about a creature near north Tumbulgum and from the descriptions he had been given, it could only be one thing.
>
> "I know a few people who have seen the creature. They have recalled it for me and provided sketches," he said. "Every time we get the same description. It looks like a large tiger or lion but it has cramped-up hind legs more like a marsupial. It has a thick stunted nose like a wombat and is covered in brindle [striped] fur with sulphur yellow spots. These descriptions fit the bill with the marsupial lion."
>
> Northern New South Wales environmental scientist Gary Opit, 64, is adamant a marsupial lion does exist, saying he has seen such a creature at least four times. Mr Opit, who hosts a weekly radio segment on the north coast about Australian wildlife, says his encounters with the beast have stretched as far north as Mt

Tamborine. "I first saw it in 1969 when I was working as a National Park Ranger at O'Reilly's," he said. "I got a perfect view of it and you could tell it was some type of marsupial because it had that waddling walk." Mr Opit, who grew up on the Gold Coast and studied at Griffith University, said he had seen a marsupial lion again in the Billinudgel Nature Reserve several times since 1995.

Jean-Marc Hero, an associate professor with the Griffith University School of Environment, is less convinced about the possibility a marsupial lion has survived. Prof. Hero said no physical evidence of live marsupial lions had been recovered since British settlement. "It's more likely to be a quoll or an escaped feral cat, which can get quite large," he said. "You certainly get quolls out at Springbrook so it would be possible a large one has ventured down further near Tumbulgum."

I'd never heard of a quoll. After reading more about the odd, carnivorous Australian marsupial (four species of it are found in Australia, two in New Guninea), it seems plausible that if people are reporting big, meat-eating cats with sulphur-colored spots, they are more than likely seeing a quoll, not a Queensland tiger.

Some claim that the Queensland tiger is merely a mainland relative of the Tasmanian tiger. That would account for the close similarities in descriptions between the two. The *Thylacoleo* became extinct in the late Pleistocene era, the thylacine or Tasmanian tiger in 1936 with the death of Benjamin in the zoo in Hobart, although the species wasn't officially declared extinct until 1986. If you had to place a bet, you'd probably go with something that wasn't officially declared extinct until thirty-odd years ago as opposed to something that became extinct thirty thousand years ago, wouldn't you? Me, too.

Other opinions expressed say one of the species of tree kangaroos, probably the larger Bennett's variety, is being mistaken for the non-arboreal thylacine, since tree kangaroos can walk on all fours on land. According to *The Age*, July 2003, "Top End Cat" describes multiple sightings of some large and strange "feline" near Cape York, Australia. There is an urban legend that American soldiers released pumas into the wild, pumas they had taken as kittens and found a lot less easy to love or care for when the cats became full-grown. Some reported seeing mountain lions, some reported Tasmanian tigers (they call them "Tassies"), and others are convinced that witnesses are actually seeing tree kangaroos.

According to *The Age*:

> Cooktown resident Joe Meaney also saw a puma-like creature leap across the road while driving into town with his wife several years ago. "If I hadn't seen it I wouldn't have believed it, I would have thought it was just an old wives' tale," he said.
>
> However, naturalist Rob Whiston from nearby Bloomfield believes the mysterious animal is a tree-climbing kangaroo. There are two main species of tree kangaroo found in north Queensland, the small blackish-brown Lumholtz and the larger Bennett's. "They have rounded ears, a hunched posture, they don't bound like other kangaroos and a beautiful tail is always streaming out behind them," he said. "They really do look quite cat-like."
>
> Tree kangaroo expert Roger Martin agrees: "Some of the old tin scratchers have told me they have seen tigers around the bush. Tree kangaroos are shy animals, they spend most of their time in the canopy and when people see them they don't know what they are." But Mr Meaney said he had a Bennett's tree kangaroo as a pet when he was young and knows the difference between a kangaroo and a cat.

Who the old "tin scratchers" might be is anyone's guess.

As I was driving through the Arizona desert very late one night, I saw a puma cross the road just a few feet in front of my car. I'll never forget it. It looked *nothing* like a tree kangaroo. It's hard to imagine that all the reports dating back hundreds, even thousands, of years by people living in the region have been merely sightings of tree kangaroos. Those aren't striped, aren't large, and they mostly stay in the tree canopy. Is the yarri, the Queensland tiger, real? We'd better hope that if they *are* real, they stay well hidden. If the fate suffered by the Tasmanian tiger is any indication, that's the only way they just might manage to survive.

CHAPTER NINETEEN

The Megalodon

Off the coast of South Africa, sea lions gather in the water in large groups known as "rafts." These pinnipeds are sometimes called Cape fur seals, but their external ears indicate otherwise. Where there are sea lions, there are the sharks that come to eat them. Still, people venture forth in small boats to fish as they always have done. They have no choice. Fishing is a way of life and has been for generations as far back as memory reaches. The boat is very old, and so, too, is the man who rows it. He knows well to avoid the active hunting ground where white death lurks beneath the surface of the waves. The massive sharks propel their bodies upward with incredible force, launching themselves and the unfortunate sea lion pups they hunt high into the air before splashing back down to savagely devour their prey.

Somehow, the sea feels different today and that makes him uneasy. The eager, ravenous, great white sharks have all disappeared as have the pups they feasted upon. The man is still within sight of land, but he is no longer close enough to make out familiar landmarks on the distant shore. The sun is going down in the west, the sky fading to pale violet, as Venus, the bright evening star, rises on the horizon. Gazing in disappointment at the meager catch still thumping and gasping in the net on the floor of his little boat, he must be philosophical. Today, the fishing was not good. Tomorrow will be better. He reluctantly points the nose of the boat back toward his coastal village and begins to row. Still, the unsettling feeling of dread persists.

From below, the sea explodes violently beneath him. The fragile wood under his feet splinters, falls away, and disappears entirely. The few fish he managed to catch gratefully escape. He feels himself being carried upward in a violent *whoosh* of foam, just like one of the unfortunate little seal lion pups he had observed so many times before. Adrenaline floods his system before he has time to form a conscious thought, and that's a mercy. He sees nothing, hears nothing. It's as if time itself pauses. Jaws as wide as his house open below him and thrust forward. They are studded thickly with row upon row of serrated, pointed, white triangles. Dull, black eyes roll back and disappear behind nictitating membranes to avoid injury. The man feels nothing but surprise. He is firmly grasped around the chest, shaken wildly like a rag doll, and gulped down into the stinking gullet of the hellish beast before he has time to take another breath. A red stain briefly marks the surface, along with a few smashed pieces of wood.

Nothing was ever found of the old man or his small fishing craft. He became simply one more among the many who ventured forth upon the indifferent sea and never returned.

There have been no tell-tale teeth washed into the shallows, other than those from *Cardaradon megalodon* who went extinct millions of years ago. There have been no verified sightings of such creatures. No modern whales have floated to the surface displaying wounds with a bite radius to indicate predation by a sixty-foot-long shark (a school bus is typically forty-five feet long). Still, the legends persist. Is a relict population of the greatest shark to ever live still breeding, reproducing, and hiding in the deepest, unexplored recesses of the oceans?

The Discovery Channel stirred up a tremendous amount of interest in the topic when it aired a documentary called "Megalodon, the Monster Shark Lives" during Shark Week in August 2013. It featured a "long-lost photo" of a World War II German U-boat with a shark dorsal and tail fin visible just behind it. Estimating the size of the shark in comparison with the sub indicated that the shark would have been monstrous. The distance between the two fins was some sixty-three feet. Ker Than, a freelance science writer who deals with topics like this one, wrote an article for *National Geographic* about the controversy that was published on August 13, 2013. For purposes of comparison, "a great white is about the size of the clasper, or penis, of a male megalodon," Peter Klimley, a shark expert at the University of California at Davis, said in a 2008 interview conducted by Than. Even more impressive is the fact that male sharks have *two* of them. A clasper isn't *exactly* a penis; it's an external organ to channel shark semen into the female during mating, but the analogy is nonetheless a powerful one.

The Discovery Channel defended itself by noting at the end of the program that "certain events and characters in this film have been dramatized" and that while "legends of giant sharks persist all over the world, there is still debate about what they may be." According to Breeanna Hare, CNN, on Friday, August 9, 2013: "In the 26-year history of Shark Week, 'Megalodon' is the highest-rated and most-watched Shark Week episode to date." Executive Producer Michael Sorenson said in the interview with Hare:

> With a whole week of Shark Week programming ahead of us, we wanted to explore the possibilities of Megalodon. It's one of the most debated shark discussions of all time, "Can Megalodon exist today?" It's ultimate Shark Week fantasy. The stories have been out there for years and with 95 percent of the ocean unexplored, who really knows?

A small disclaimer: the scientists were actors, the so-called evidence faked. This docufiction/mockumentary inspired a good deal of indignant backlash from scientists and viewers alike.

Megalodon is believed to have lived between 23 (although some sources put it at 28) and 2.6 million years ago, beginning in the early Miocene and becoming extinct during the Pliocene era when seas were considerably warmer than they are now. It was named *Carcaradon megalodon* in 1835 by a Swiss naturalist, Louis Agassiz. The megalodon was about three times as big as the largest known great white shark, *Carcaradon carcarias*. Twenty feet is believed to be a maximum length for great whites, and one reaching that size would be exceedingly rare. Megalodon teeth that continue to turn up can be as long as seven-and-three-eighths inches! I have one just a shade under six inches, almost as long as my hand, and very sharply serrated; take my word, you would not want a mouthful of them sinking into you. A very big great white's tooth is less than three inches. Pliny the Elder of Italy, the same one who documented the eruption of Mt. Vesuvius in A.D. 79, was the first writer to mention shark teeth fossils, but he incorrectly identified them, mistakenly believing they fell to earth during lunar eclipses. So large and fearsome were the teeth, in fact, that people during the time of the Renaissance thought megalodon teeth must have been the fossilized tongues of dragons or snakes. They called them tongue stones or *glossopetrae.* Royals carried them for luck. They were ground up to use medicinally as anti-toxins.

In 1858, Ebenezer Emmons, in his "Report of the North-Carolina Geological Survey; Agriculture of the Eastern Counties: Together with Descriptions of the Fossils of the Marl Beds," wrote:

> If the size of the teeth furnish an indication of the strength, size and ferocity of this species of shark, then it must have been one of the largest and most formidable animals of the ocean, combining, as Prof. Owen remarks, with the organization of the shark, its bold and insatiable character, they must have constituted the most terrific and irresistible of the predaceous monsters of the ancient deep. The largest of the teeth measure sometimes six inches in length, and from four to five wide at base.

In fact, the megalodon tooth is the state fossil of North Carolina.

For years, based on similarities between their teeth, people have assumed that megalodons were simply a sort of overgrown precursor to the modern great whites. This hypothesis has now been called into question and is no longer accepted as fact. Some researchers believe the two species evolved along similar lines, but that modern great whites are not the descendants of ancient megalodons. The two separate species overlapped for a period of approximately ten million years. While great whites may opportunistically scavenge whale carcasses, evidence in the fossil record shows megalodons actually attacked living whales. They may have bitten off their fins making locomotion impossible and then devoured them at leisure. Such an apex predator is almost beyond imagining.

The summer *Jaws* was released, 1975, I saw two great white sharks that had been caught off the California coast and frozen in glass cases put on display at Sea World in San Diego. In the stomach of one of them, a 180-pound sea lion was found. It had been swallowed whole. For a person to imagine the kind of animal who could pull off such a feat is terrifying enough. To imagine a shark three times that powerful, three times that enormous, is nearly impossible.

Megalodon lived throughout the oceans of the world as evidenced by their teeth being discovered in widely diverse locales, but the seas were far warmer than they are today. The descriptor "cosmopolitan" appears in just about every published source dealing with their lifestyle; so often, in fact, that you are quite tempted to imagine them sipping martinis while jetting off, first class, to a new locale. The last of them are believed to have died out during the Pleistocene extinction when large sheets of glacial ice covered the continents and the oceans cooled considerably. This occurred at the same time that megafauna, the great mammals, also vanished. While we don't expect to come upon a saber-toothed cat or wooly rhino, plenty of people think that a huge shark still lurking in the deepest part of the ocean is entirely possible.

Unless the species completely altered its pattern of behavior, something other species were unable to do, the ocean depths would be the last place megalodon would have chosen to live. It was not a pelagic (deep, open water) shark. Its prey was mostly whales and dolphins, cetaceans that lived in shallower waters, too, and certainly never ventured into the coldest depths. Megalodons also preferred warm, shallow coastal bays and inlets to give birth and nurse their pups. While other, smaller sharks adapted to changes in the ocean temperature, the largest among them didn't seem able to do so. Conserving heat in so large a creature would have posed a formidable challenge. They would have been forced to make do in increasingly smaller and smaller shrinking bodies of water warm enough to support them. Their pups would have been subject to increased predation. There would have been competition from other animals such as orcas. Like great whites, they gave birth to relatively few young and their pups took a long time to reach sexual maturity. The changes in climate and increasing competition from other species ultimately proved too much for megalodon to overcome.

Most life that occurs in the oceans is found in places where sunlight can penetrate. The top layer, about 650 feet, is called the euphotic zone. The second level, the dysphotic zone, reaches down to about 3,300 feet, and at those depths, photosynthesis becomes impossible. It is dark but still has some light penetration. The final layer of the ocean is entirely without light and is called aphotic. Bioluminescent creatures dwell there in the darkness, along with the giant squids. We have evidence of the giant squids, however. Their remains are found in the stomachs of sperm whales. Their bodies occasionally wash up on shore. Films have at last been made of them swimming in the awesome depths. No similar evidence exists indicating the existence of a giant species of shark.

In 1976, an entirely new species of shark was discovered. It was a large species, too. Along with that of the coelacanth, its discovery was one of the most notable marine-life finds of the twentieth century. The megamouth shark, *Megachasma pelagios*, had been there all along, unbeknownst to us, but no one had suspected, captured, or filmed one. The eighteen-foot-long animals are large filter-feeders, like the whale shark and basking shark, who swim deeper during the day than at night. They follow the plankton as it rises. The first specimen ever caught became tangled in the sea anchor line of a US Navy ship off Kāne'ohe, Hawaii. The startling fact of the megamouth's existence is proof positive that we have not yet discovered all there is to know about life in the sea.

Are monstrous sharks sharing our world? Could sharp-toothed sharks the length of bowling lanes, the width of tennis courts, be going about the business of survival in the deep, dark trenches of the cold oceans of the world without anyone being aware of it? Sure, it's possible. Is it likely? Not so much.

CHAPTER TWENTY

The Mothman

In a small town like Point Pleasant, really just a blip on the map in West Virginia where the Ohio and Kanawah Rivers come together, it was difficult to evade the prying eyes of your parents, your neighbors, and even your friends. There were not even six-thousand residents, but you'd swear every one of them had at least six eyes apiece, judging from the amount of gossip exchanged. Sometimes, it felt like everyone knew everyone else's business, and if they *didn't* already know it they were busy trying to uncover it. Figuring out how to get some time alone with your best girl required more study than the heinous trig exam Mr. Norton concocted and administered last week.

"So, where we going, Jason?" Kirstin asked her boyfriend, sliding across the bench seat of his nine-year-old '57 Chevy, his pride and joy, to rest her head on his broad shoulder. "Not like there's a lot of choices." She grinned wryly. It was the familiar lament of Point Pleasant's teenaged residents.

"I'll give you one clue." He grinned back. "It's explosive! *BOOM!*" He playfully grabbed his girlfriend as he said the last word, and she giggled appreciatively.

There wasn't really anywhere else near town where kids could go to hang out, or more precisely, to *make* out, except the old TNT munitions plant. It was creepy and deserted and hadn't seen any action since the end of World War II, but if the way Kirstin was nibbling on his earlobe on the drive out there was any indication, it was sure gonna see some action tonight. Whooo-boy.

He parked the car near the dark stand of woods at the back of the parking lot and turned to kiss her. "You look totally bitchin', babe," he said, brushing a stray strand of long hair out of her eyes. She wore hip-hugging bell bottomed jeans and the thin, ribbed shirt known as a Poor Boy. It showed off her figure to advantage.

The sun was setting and the evening was already turning chilly. She shivered slightly.

"Here, take my varsity jacket," he offered generously. He couldn't wrestle out of it very easily behind the wheel, he was a big guy, so he stepped out of the car for just a second and shrugged it off.

"Uh, Kirstin? You see that?" he whispered urgently, passing her his letterman's jacket, black with red trim and the image of a scowling knight in armor on the front. He nodded slowly toward the line of trees a few yards in front of the hood of his Chevy. His brown eyes were wide and getting wider.

The last rays of the setting sun glinted off two round, red eyes in the foliage of the trees. They seemed to be peering right at the two teenagers. The figure behind them was gray-brown with a rounded head. Suddenly, as Jason jumped back into the Chevy, it took flight and glided from the branch where it had perched and swooped swiftly and silently toward the roof of the car. Its long legs ended in wickedly sharp talons.

"*Aaaaggghhh!*" Kirstin let out a high-pitched squeal and buried her pretty face against Jason's well-muscled chest. "Let's get outta here—*now. Go!*"

"Okay, okay, I'm going," he agreed, fumbling for the key and shoving it into the ignition. He stomped down hard on the gas pedal and peeled out of the cracked, weedy parking lot, leaving a cloud of dust in his wake.

"Is it gone?" she whimpered, looking around apprehensively through the car windows.

"Yeah, I think so."

"Good because...*eeeeeee!*"

"Oh my *God*, it's following us!" Jason yelled.

Just then, a large, dark figure swooped past the front windshield in a rush of wings and feathers. Its glowing red eyes turned briefly to stare directly into those of the terrified young couple, and then the creature rose straight up into the sky and disappeared.

They looked at each other and said at exactly the same moment, "Mothman!"

While reports of some strange beasts reach back through the veil of time hundreds, even thousands of years, the mothman's appearance in West Virginia is a relatively recent and well-documented phenomena. Like others of its kind, though, there have also been hoaxes. Every time there is a sincere desire on the part of a witness to report a sighting of some bizarre beast, there are those who gleefully jump on the bandwagon for financial or personal gain.

The first Mothman sighting occurred on November 12, 1966. It happened in a graveyard. That alone ramps up the weird factor and causes skeptics to nod their heads knowingly. Five men were digging a grave in Clendenin, West Virginia, when a creature that looked like a flying man swooped low over their heads from the trees nearby and flew away. The figure apparently and understandably scared the bejesus out of the gravediggers—who presumably, based upon their occupation, don't scare all that easily.

The West Virginia Commerce website proudly, almost breathlessly touts the exciting event this way:

> November 12, 1966: On this date, five men digging a grave in a cemetery near Clendenin, West Virginia, saw something that looked like "a brown human being" that flew from some nearby trees and glided low over their heads. This was the first sighting of the Mothman. There is now a Mothman Museum as well as an annual Mothman Festival in September. This was also the basis for the movie *Mothman Prophecies* staring Richard Gere.
>
> Location: Point Pleasant, WVa., County: Mason County.

You can almost see the local politicians and city fathers rubbing their hands cheerfully together imagining the windfall of tourist dollars flowing into public coffers. Museum! Festival! Richard Gere! Yes, there's certainly some very easy money to be made from the Mothman. There is also a larger-than-life-sized twelve-foot-tall statue of him in town created by Bob Roach and unveiled in 2003. It makes Mothman look like the largest, ugliest butterfly you can imagine, but hey, it still pulled a very respectable 4.5 stars on TripAdvisor. The possibilities seem almost endless, don't they?

An interesting side note which throws some light on the Mothman phenomena is a similar incident, one that happened fourteen years earlier in West Virginia. There was a creature in the trees referred to as the Flatwoods monster. The flying monster was widely reputed to be an alien. It was first sighted and reported in September 1952. It had a green, pleated sort of exoskeleton, a head shaped like an ace of spaces, and two small arms that ended in claws that were held out in front of it. It had red, glowing eyes, and it made odd sounds. After much study and speculation, most people now are convinced that the monster was very likely a barn owl. The green, "pleated" body was probably the leaves and foliage of the tree in which it roosted. That fact made it look much taller than it actually was. The arms with "claws" would have been the legs and talons of the owl gripping a branch. A meteor sighting at the same time added to the other-worldly nature of the encounter. The Flatwoods monster hasn't been forgotten—there is a three-day music festival in its honor every year.

A few days after the five gravediggers' puzzling sighting in Point Pleasant, five young people reported that they, too, saw the Mothman on November 16, but within the span of four days between the first and second reports, the story had expanded. This time, the kids were hanging out near the old, abandoned World War II TNT factory outside of town. Mothman approached them. He was seven feet tall. He had red, glowing eyes. Because of the height and the fact that red appears on its head, a sand hill crane who got lost while migrating is sometimes put forth as a possible, and somewhat unlikely, candidate for Mothman. Although the kids drove at speeds up to one hundred miles an hour in an effort to escape, Mothman had no trouble keeping up with their car as they fled toward the sheriff's office in town.

Deputy Millard Halstad had known the eyewitnesses, two couples and one of their cousins, all their lives and believed the strange story. So did the many others in town who continued to report sightings over the next year. One woman was so startled by the sight of the red-eyed Mothman that she dropped her baby daughter and fell on top of the girl, paralyzed with fear. Everyone who saw the Mothman said he didn't flap his wings as a bird would do. He simply opened his wings and ascended like a hot-air balloon into the air. There were around one-hundred reports of encounters with the Mothman between November 1966 and November 1967.

What if Mothman was, like the Flatwoods monster fourteen years earlier, just an owl? Owl feathers are unusual. They are designed to be virtually soundless in flight. This allows the predator to ambush small animals upon which it feeds without alerting them to its presence until the owl is only inches away, and then it's too late. Wings from most birds make flapping or whooshing sounds as they fly. Not those of the owl. Owl feathers have a series of very tiny cuts along the edges that absorb sound. It's a strange sensation to watch an owl in flight because you expect to hear sounds as you would with any other bird, but you don't. You hear nothing.

According to an article in *Strategy* about owls, published on April 16, 2016:

> Wing feathers enable near-silent flight. First, the leading edge of the owl's wing has feathers covered in small structures (hooks and bows) that break up the flowing air into smaller, micro-turbulences. These smaller areas of turbulence then roll along the owl's wing toward the trailing edge, which is comprised of a flexible fringe. This fringe breaks up the air further as it flows off the trailing edge, resulting in a large reduction in aerodynamic noise. Then, any remaining noise that would be detectable by the owl's prey is absorbed by velvety down feathers on the owl's wings and legs. These soft feathers absorb high frequency sounds that most prey, as well as humans, are sensitive to. All together, these features enable owls to remain undetected when they fly. However, it's believed that the wing's serrated leading edge is most effective at reducing noise when the wing is at a steep angle—which would happen when the owl is close to its prey and coming in for a strike.

It's not surprising Mothman made no sound in flight, because he is likely an owl.

People report Mothman's red, glowing eyes. *Tapetum lucidum*, a layer of tissue, reflects visible light from the eye itself back to the source of the light. It accounts for the color of eyes shining in the dark. Animals' eyes in the darkness when they are struck by a light source have different colors of what is sometimes called "eyeshine." When you take photos of people and later notice that there is "red eye" on your subjects, that is the blood-rich fundus at the back of the retina showing through the opening of the pupil behind the clear cornea. Humans don't have a *tapetum lucidum*, but animals that hunt or are active in the dark do. That accounts for their superior night vision. Dogs and cats usually have green or yellowish eyeshine, white-tailed deer have white, horses have blue, and many kinds of birds and *owls* have eyes that shine red in low light. If Mothman were a man, he would not have eyeshine at all. The eyeshine of an owl is red. In the same way some people see a large, unusual bird and believe it is a thunderbird, others see a big, red-eyed owl near nightfall and call it Mothman.

Shortly after the first sightings were reported, a tragedy occurred which added a dimension of prophecy to the Mothman's appearance. The Silver Bridge, some seven-hundred-feet long, over the Ohio River connecting Ohio and West Virginia collapsed on December 15, 1967. It happened at about 5:00 PM during rush hour. Forty-six people were killed in the tragedy, and two of the bodies were never recovered. Was Mothman trying to warn citizens in the area of the impending collapse, people wondered? Was mothman the *cause* of the lethal accident? Speculation ran rampant, speculation that linked the appearance of Mothman with the disaster. The National Institute of Standards and Technology Museum has in its collection the I-bar from the Silver Bridge that fractured, the same cracked I-bar that was responsible for the collapse.

There is also on display a scale model of the ill-fated bridge with the following description:

> On December 15, 1967, the Silver Bridge connecting Point Pleasant, West Virginia, with Kanauga, Ohio, collapsed into the Ohio River sending 46 people to their deaths. Scientists from the National Bureau of Standards (NBS) were called upon by the National Transportation Safety Board to investigate the cause of the disaster. NBS was selected because of its competence in metallurgy and its expertise in failure analysis. This working scale model of the Silver Bridge was constructed by the NBS Metallurgy Division and used in the investigation of the disaster. NBS submitted its findings to the task force on bridge safety appointed by the President. The task force concluded that a small crack had developed at a small corrosion pit on eyebar 330 and had grown to a critical size of 1/8 inch by the joint action of sulfur-based stress corrosion cracking and corrosion fatigue. This was cited as the cause of the Silver Bridge collapse. (See also artifact 2010.0098.001.)

Nowhere is the mothman's influence mentioned.

One summer evening, just as the sun was dipping behind the woods that surround my house, I heard the characteristic call of an owl. It sounded a little like it was asking the old "who-cooks-for-you?" question. I couldn't help myself. I called back, low and trilling and as realistic as I could make it. Houses in my neighborhood are all at least a couple of acres away from each other, so I didn't worry too much about people thinking I'd lost my mind. Of course I didn't ask who the owl's chef was. It was more a "whoo-whoo-whoo-*WHOOOO*?" repeated twice. To my astonishment, the owl answered back. Not only that, from deep in the woods, it started to approach me! I could see its silent, shadowy form flitting through the canopy of trees toward my back deck. I kept up my end of the conversation. From reading Jane Yolen's *Owl Moon* to my children many years ago, I knew owls could be called if you were good enough at mimicking their call. Apparently, I was.

Soon, I saw a truly enormous gray-brown form rounded on top with wide, round, dark eyes outlined in concentric circles of pale gray, bobbing its head up and down and weaving rhythmically back and forth on the branch of a wild cherry tree. Since the sun was behind the owl, the eyes didn't glow red. We called companionably to each other for a few minutes. Finally, the owl realized that the odd creature sitting on the deck sipping iced tea was neither a potential mate nor a territorial rival after all, regardless of what it might sound like. After the flummoxed bird disappeared back into the forest, I looked it up in my trusty *Field Guide to Birds of North America*. I'd seen a great horned owl before in the neighborhood, and once, while driving across Wyoming, a snowy owl flew right beside my car for a hundred yards or so. A tiny screech owl no taller than eight inches would roost on my front pediment porch every night for years.

This loquacious visitor was a barred owl, *Strix varia,* one of the largest owls in North America. Its range is the entire eastern half of the county, well into Canada, and even the coastal mountains of Mexico. It stands up to twenty-five inches tall with a wingspan of up to fifty inches. Watch the opening scene of Harry Potter, the first one, and you'll see a tremendously large owl, most probably a great gray owl, quite similar to the barred owl I called into my backyard. They are genuinely impressive. I would have had to consume a prodigious amount of alcohol, however, and not merely iced tea, before I'd have claimed that the big bird I saw that evening was a Mothman.

CHAPTER TWENTY-ONE

The Wendigo

Cold seeped into every seam that didn't quite meet, every warped window and door frame, every tiny chink in the mud packed between the rough-hewn logs of the one-room log cabin. Snow sifted in along with the cold, too, until there seemed to be little difference between inside and outside. Food was a distant memory, fast fading in his mind. All he felt now was emptiness, a hollow in his belly that would never, could never, be filled. The fire that crackled merrily in the hearth mocked his despair. He sat in front of it now, staring blankly into the dancing flames. The end wasn't far off, and he knew this, knew it with the same certainty that he knew the whole, wretched scheme had been a fool's dream. Trapping furs in the rugged north country, fat profits there for the taking, had seemed like a brilliant idea last spring. It would be a romantic adventure, one that would make him rich.

He'd felt incredibly lucky to find an abandoned cabin and used it as his home base through the fall and this unending winter. Surely it would end, wouldn't it? It hadn't occurred to him to question what had become of the home's previous resident. He moved right in and made it his own. Plush pelts, fleshed and dried and tanned and ready to trade, were stacked neatly in piles. He'd tried boiling a few and chewing on the skin, but that was useless. It only sharpened his hunger. The animals themselves, once so plentiful, were nowhere to be found; they nestled deep in their underground burrows with stores of acorns and seeds to keep them alive until spring. Spring. He wouldn't live to see it.

There came a knock at the door.

"Anyone there?" called a muffled voice over the howling of the bitter wind.

"Yes!" he cried. "Oh, God, yes!" At last he would be delivered from starvation. Help had arrived!

He flung wide the door with a grateful smile. At once, his face fell. Relief faded swiftly to resignation. The stranger on his doorstep looked every bit as hungry as he was himself. He carried almost nothing and looked about spent. There was barely any meat on his bones at all.

"Come in."

"Thank goodness," the stranger gasped, falling heavily into the room and collapsing on the splintered planks of the floor. "Fire. Food."

"Well, fire anyway," the man said quietly.

The stranger looked at the empty pot near the hearth, the bare table, the shelves that held no provisions. Then he looked at his host. The man who looked back at him was gaunt, gray, and filthy. Lank, greasy hair hung to his shoulders. Every bone was visible in the sharp planes of his face. Prominent eye sockets, the brow ridge, sunken cheeks with pointed cheekbones made what remained of his face look like a skull. Worn clothes hung upon him like the tattered clothes of a scarecrow. The only part of him that seemed alive were bright, glittering eyes. The man smiled at his guest.

"You're just in time."

"Just in time for what," the stranger queried uneasily.

"Supper."

Growing up on the West Coast and living much of my adult life in Iowa City, I hadn't heard of Wendigos until a few years ago when one appeared on an episode of *Haven* on the SyFy channel. The basic concept is easy enough to understand. In times of famine and privation, people will be sorely tempted to eat anything remotely edible—including friends and family. Remember the Chuck Jones-directed Bugs Bunny cartoon from 1943, *Wackiki Wabbit*, where two starving castaways are marooned on a desert island? One is short and rotund, the other tall and thin. As they are about to cook Bugs, all the while gleefully chanting, "We're gonna have roast rabbit," the elusive bunny boards a cruise ship and departs. As starvation sets in, they appear to resemble a tempting hamburger and a hot dog—with legs. The finale? They run off into the distance trying to catch and eat each other.

Even in this country, people are sometimes hungry, but there are bureaucratic safeguards in place to prevent actual starvation on a widespread scale. We see those heartbreaking photos, of course, on the nightly news or in magazines, but the people with stick-like arms and legs and painfully distended abdomens aren't usually part of our personal experience. Families in this country are eligible for a Supplemental Nutritional Assistance Program card to purchase food. Schools provide breakfast and lunch for children who need them. There are also community food pantries and free meals distributed with no questions asked by a variety of charitable organizations. Hunger is one thing, starvation quite another. Before safety nets like these types of assistance programs existed, however, one failed harvest, one miscalculation when following an unfamiliar trail, or one uncommonly long, fierce winter could result in swift disaster—starvation!

The bones of a few of our ancestors at the ill-fated Jamestown settlement showed evidence of having been gnawed upon, and *not* by animals. Most of us, especially those who studied California history in fourth grade, learned at some point about the hapless Donner party trapped during the winter by drifts of deep snow in the Sierra Nevada mountains in 1846–1847. Some members of the little group heading westward were forced to consume the unlucky others who had already died in order to ensure their own survival. The infamous pass there bears their name and still has the ability to bring a shudder some 170 years later. That is the same macabre fascination people have with *Alive*, the saga of Uruguay's rugby team whose plane crashed in the Andes. Cannibalism remains one of humanity's most sacrosanct taboos. To break it, people must be driven to the edge, either desperate with hunger or mentally disturbed. Think Hannibal Lecter or Jeffrey Dahmer as examples of the latter.

The legend of the Wendigo was born among the Algonquian people, the Cree, Ojibway, Salteaux, Naskapi, and others, during those long-ago times when the long shadow of famine stalked the frozen northland, and people feared the terrible, irresistible, insatiable desire to consume *anything* edible. The legend took root in the forested regions of the North Atlantic states and Canada and in areas surrounding the Great Lakes. Not only the eating of human flesh but any unmitigated greed or compelling lust for killing could be indications that a Wendigo spirit had taken possession of a person. (The name of the spirit has various spellings, depending upon the source. Unless a particular source spells it otherwise, Wendigo is the spelling I use.)

Ojibway storyteller Basil Johnston, in *The Manitous, the Super-natural World of the Ojibway* (1996), describes Wendigos as spirits who "came into being in winter and stopped villagers and beset wanderers. Ever hungry, they craved human flesh, which is the only substance that could sustain them. The irony is that having eaten human flesh, the windigoes grew in size, so their hunger and craving remained in proportion to their size; thus they were eternally starving." The spirit Johnson describes is terribly thin and ash-gray with eyes sunken into their sockets and flesh covered in weeping sores. It reeks of putrid decay.

In times of severe want, the ceremony of the Wendigo was sometimes performed in order to remind people to be wary of possession by the dangerous spirit. It reinforced the taboo about eating human flesh. A lake in Minnesota, on the Leech Lake Indian Reservation, bears the name Windigo. The lake is located on Star Island in the middle of Cass Lake, making it an extremely rare spring-fed lake within a lake. This was the place where the last known ceremony of the Wendigo was performed in the United States, but the date of this event is unclear. In the early part of the twentieth century, small parts of Star Island were sold to private individuals, but most of it remains a national forest preserve. William Warren, another member of the Ojibway, writes about this ceremony in *History of the Ojibway People* (1994).

In *American Anthropologist*, August 1960, researcher Seymour Parker writes about the mental illness widely known as Wendigo in "The Wiitiko Psychosis in the Context of Ojibwa Personality and Culture." He calls it a "bizarre form of mental disorder involving obsessive cannibalism." An afflicted individual "begins to see those around him (often family members) as fat, luscious animals, which he desires to devour." If the illness progresses beyond this point, it can result in "violent homicidal cannibalism."

When this occurs, Parker writes, there is no other option in the culture than putting the victim of the psychosis to death, although other researchers within the Algonquian-speaking culture disagree that death is the only way to stop a Wendigo. When there exists an objectively realistic and ever-present fear of starvation among a group of people, as there was among the Ojibway who suffered periodic food scarcity, contracting the wendigo psychosis was a potential result.

Gord Bruyere, in *Spiritualty and Social Work, Selected Canadian Readings* (2007), writes:

> The Windigo were once human beings who lived among us as family and contributed to our society as much as any other person or clan. Yet through some agony endured, those human beings were made to eat the flesh of other human beings. Sometimes it was overwhelming hunger brought on by the harshest of winters that caused this to happen. In other instances, people were told to do this through the force of their dreams.

Bruyere says that there were ways to cure a person taken by the Wendigo sickness, but such an intervention required "the will of the entire community" and the combined efforts of healers, elders, and medicine people. It was "a most delicate spiritual matter." Existing in balance and harmony within the community was the goal of all such ceremonies.

A Cree trapper named Swift Runner who lived in a campsite near Edmonton, Alberta, Canada, was thought to be a victim of the Wendigo sickness. The year was 1878, and his family was starving. Even though he was within a day's travel from the Hudson's Bay Trading Company where food was available, Swift Runner first ate his oldest son who had died of starvation and then killed and ate his wife and their five remaining children. Over the course of the long winter, he also killed and ate his mother and brother. When he showed up at a church in the spring, priests were skeptical of his account. He weighed about 200 pounds and did not appear to be starving. Andrew Hanon, a reporter for the *Sun Times*, wrote about the infamous murders on August 12, 2008, in an article titled "Evil Spirit Makes Man Eat Family." Below, Hanon interviews Nathan Carlson, a First Nations writer, leading expert on the Wendigo, and, as he discovered while conducting his research, a distant relative of the man who ate his family:

> Swift Runner was hanged in (December) 1879 in Fort Saskatchewan, the first legal execution in Alberta. The macabre case is considered by many to be the most horrifying crime in the province's history.

> But what most people don't realize is that it was part of a much larger phenomenon that Edmonton ethno-historian Nathan Carlson calls Windigo condition, which haunted communities right across northern Alberta in the late 19th and early 20th centuries and cost dozens of lives.

The Windigo (an Anglicized form of the word Witiko) is a mythological creature among native cultures from the Rockies to northern Quebec. It has an insatiable appetite for human flesh and wreaks destruction wherever it goes.

Carlson describes it as "the consummate predator of humanity." It's sometimes described as "an owl-eyed monster with large claws, matted hair, a naked emaciated body and a heart made of solid ice."

"It's extremely destructive," he says. "The more it eats, the hungrier it gets, so it just keeps killing."

Windigos can possess people, transforming them into wild-eyed, violent, flesh-eating maniacs with superhuman strength. Many native people in northern Alberta lived in terror of being possessed.

During the trial, records revealed that the killer "eventually confessed that he shot some of his family, bludgeoned others with an axe and even strangled one girl with a cord. In some accounts, Swift Runner said he fed one boy human flesh before he too was killed."

Another case involved Jack Fiddler and his brother, Joseph, First Nations people from Canada. In 1907, the two men were tried and convicted for the murder of a woman they had killed, claiming she was possessed by a Wendigo spirit. After the brothers were incarcerated, Jack committed suicide. Joseph died in prison three days before he was to have been pardoned.

Algernon Blackwood wrote a horror story called "The Wendigo" in 1910. For many, this tale was their first introduction to the fearsome Algonquian spirit monster. Because of Project Gutenberg, the story is available free online with very few restrictions for anyone to read. It's dramatically written in the manly man versus nature spirit of Jack London and suitably evokes the creepy Wendigo that has possessed an unfortunate character named Defago:

> Nothing really can describe that ghastly caricature, that parody, masquerading there in the firelight as Defago. From the ruins of the dark and awful memories he still retains, Simpson declares that the face was more animal than human, the features drawn about into wrong proportions, the skin loose and hanging, as though he had been subjected to extraordinary pressures and tensions. It made him think vaguely of those bladder faces blown

up by the hawkers on Ludgate Hill, that change their expression as they swell, and as they collapse emit a faint and wailing imitation of a voice.

The story ends with this final chilling line: "Defago had 'seen the Wendigo.'"

Modern television programs like *Sleepy Hollow*, *Teen Wolf*, *Charmed*, *Supernatural*, and *Haven* each have episodes about the Wendigo. "Wendigo" is the name of a character in the Marvel Comics Universe. Stephen King includes one in his novel *Pet Sematary*. The evil, emaciated spirits appear in Dungeons and Dragons and video games. These once-obscure creatures of Algonquian legend are now firmly and forever a part of pop culture.

The next time you miss a meal and feel the beginnings of hunger pangs, be relieved that you live in the land of plenty, the land "above the fruited plain" where all of those carefully cultivated "amber waves of grain" keep the Wendigo away from *your* door.

CHAPTER TWENTY-TWO

Hype, Hokum, and Historical Hoaxes

No one would love to see one—or more—of the fabulous creatures of legend (or one not yet even imagined) proven real. Since I was very small, I've hoped against hope that someone somewhere would find a real sea serpent, a mermaid, a Yeti, or a sasquatch. Every new report over the decades brought with it renewed hope. Maybe *this* time... In every case, however, hopes were dashed—mine and those of other dreamers like me. Do I still believe that one day, a new creature so amazing as to defy belief will reveal itself and withstand the tests of scientific scrutiny? I do. Unfortunately, apart from the genuine eye-poppers already discussed in the introduction, I simply don't believe that it has happened yet.

People, unscrupulous people, have deliberately made it very difficult to tell a genuine mysterious sighting or discovery of a hitherto unknown kind of beast from a clever (or sometimes clumsy) attempt to deliberately fool the general public. If we could take every report at face value, things would be a lot simpler when trying to figure out just exactly what kinds of strange and unusual creatures people are actually encountering. There have been many of these hoaxes perpetrated over the years, some to garner notoriety for those who create them and others for simple monetary gain. Here are a few of the most famous you may have heard about and some you probably haven't.

In the little English village of Piltdown, East Sussex, in 1908, Charles Dawson claimed to have discovered the fabled "missing link" between ape and man. In 1912, he contacted geologist Arthur Smith Woodward of England's Natural History Museum with bits of skull, a jawbone, teeth, and simple tools he said he found at a Pleistocene era gravel bed. This wasn't his first, nor would it be his last, carefully constructed plot to hoodwink the British public. Not until 1953 was the "evidence" proven once and for all to have been deliberately faked.

The skull of a medieval man who possessed a relatively smaller brain than average, the altered jaw from an orangutan, and chimpanzee teeth filed down and then held in place by dental putty were used by Dawson to shore up his claim. Experts challenged the "missing link" conclusion almost immediately, and yet after Dawson's death in 1916, the hoax only gained steam. Thousands of textbooks eventually had to be revised, while many scholars wasted months and years of scientific research and funding using the faked fossils as a basis for inquiry leading them only to unproductive dead ends. Sarah Kaplan, reporting for the *Washington Post* on August 11, 2016, says Dawson "misled scientists for four decades." Kaplan continues, "The scientists called the fake 'extraordinarily skillful,' and the hoax 'so entirely unscrupulous and inexplicable as to find no parallel in the history of paleontological discovery.'"

At least thirty-eight of his so-called "discoveries" were faked by Dawson, according to the *BBC News*, in an article, "Charles Dawson: the Piltdown Faker," published on November 21, 2003. They ranged from a petrified toad surrounded by a nodule of flint to a hybrid goldfish-carp to his report of a sea serpent merrily paddling about in the English Channel. The earlier, simpler fabrications helped to prepare him for "the big one," the greatest hoax of his career.

The *BBC News* article likens Dawson to a Dr. Jeykell/Mr. Hyde personality, that of a helpful country legal solicitor by day and an amoral liar and creator of elaborate fakes in his off hours. "Piltdown was not a 'one-off' hoax, more the culmination of a life's work." Although he died before realizing the extent of the damage he caused, Dawson's impact on British paleontology was long-lasting and highly destructive. He was far from the only notorious faker.

Basilosaurus, whose name means King Lizard, was a prehistoric whale, an actual sea monster that lived about forty million years ago. It reached lengths of up to sixty feet and had a mouthful of sharp teeth. Aren't those impressive enough credentials for basilosaurus to be respected in its own right? Not for Albert Koch, German immigrant, they weren't. Koch took the fossilized remains from five, some say six, basilosaurus skeletons and put them together to form a creature that never lived, a creature one-hundred-fifteen-feet long. He called his creation Hydrarchos. It was neither ignorance nor incompetence that caused Koch to cobble together this animal. Instead, it was the urge to earn undeserved fame and money for himself, even if he had to dupe people to do it. Koch was one of the nineteenth century's great "fossil showmen" who had people lining up around the block to see natural curiosities, some of them real, many of them fakes.

According to naturalist Jeffries Wyman, Koch used not just whale fossils but also common ammonite shells. Koch called the fictive beast *Hydrargos sillimani,* after Yale professor Benjamin Silliman, who had nothing to do with the hoax and demanded that Koch change the name. He obliged, renaming it Hyrarchos, after no one in particular. In *Alabama Heritage,* issue 12, 1989, in an article titled "Immense Antediluvian Monsters," Douglas E. Jones called the creature assembled by Dr. Koch (although Koch had earned neither an MD nor a Ph.D) "a crazy-quilt monster" made from "the bones of multiple fossil whales." A contemporary of Phineas T. Barnum, Koch took a page from America's master of deceptive hokum and lined his own pockets in the process.

It wasn't only adults who jumped on the hoaxers' bandwagon. When amateur photography was still a novelty back in 1917, a pair of English schoolgirls, cousins Elsie Wright and Frances Griffiths, hoodwinked the country with their "unretouched" photos of themselves posing with fairies in Cottingley. The entire event is beautifully documented in *FairyTale: A True Story* (1997), except that in the film, the director couldn't resist including fanciful visits from actual fairies (computer generated, of course), so that calling it "a true story" is not *quite* true after all. It is "loosely based" on the actual events. Everyone from Harry Houdini to Sir Arthur Conan Doyle took part in the scrutiny of the famous photos, Houdini to debunk them and Doyle to champion them.

I first heard about this hoax in "The Man Who Believed in Fairies," an article by Mike Huntington published in the *Smithsonian* on August 31, 1997. There were fascinating reproductions of the photos in the magazine. Huntington wrote:

In 1917, two girls from the Yorkshire village of Cottingley made photographs of themselves cavorting with fairies. Few took the pictures seriously, but Doyle did. He wrote a book defending their authenticity. "And what a joy," he enthused, "is in the complete abandon of [the fairies'] little graceful figures as they let themselves go in the dance! They may have their shadows and trials [but] there is a great gladness manifest in this demonstration of their life.".... Doyle died in 1930, still a believer.

Most know Sir Arthur Conan Doyle for his most famous literary creation, detective Sherlock Holmes. What a lot of people don't realize is that he was also an ardent believer in psychic phenomena. If you are unfamiliar with Doyle's personal history, his beloved son was killed during WWI, and he and his wife spent many years (and pounds sterling) trying to contact the young man beyond the grave. Doyle was one of the most famous adherents of the spiritualist movement, while Houdini exposed how "visits" from the spirits at the seances so popular at that time could be faked to bilk grief-stricken, gullible loved ones. Houdini vowed that if there were any way to return from beyond the grave and communicate with the world of the living, he would do so on October 31, the anniversary of his death. He never has.

The Cottingley fairies hoax was one of the earliest, most notorious, and successful attempts to dupe the public by using photographs, a hoax that helped make any and all subsequent photographic evidence suspect. Now, because of repeated deceptions like this one, even good photos are assumed to have been doctored or at least staged. (Recall the "surgeon's photo" of the Loch Ness monster, the photo that *proved* Nessie's existence? It wasn't doctored, per se, but it was a picture of a clay dino head mounted on a toy submarine.) The girls used paper cut-outs of fairies and posed them with hat pins and string in their charming garden alongside a picturesque stream, disposing of the telltale evidence afterward.

James Randi, master magician and debunker of hoaxes who is also known as The Amazing Randi, discovered that the photographs showed fairies uncannily similar (and after looking at the original book illustrations myself, I'd say they are virtually identical, except for the fact that the girls enlarged the fairy wings) to those found in the children's book, *Princess Mary's Gift Book*. The illustrated volume was published by London's Hodder and

Stoughton in 1915, shortly before the first photographs were taken in Cottingley. The two girls steadfastly maintained the authenticity of the pictures until 1983 when they finally admitted their deception—but at the same time, they maintained, they had actually *seen* genuine fairies.

By then, of course, their ludicrous claim wouldn't hold water. Francis died in 1986 and Elsie in 1988. Most people can sympathize with young girls getting in *way* over their heads by spinning a web of lies. It happens. Staunchly defending their fake fairies for sixty-six years and then ultimately admitting their photographic trickery while still hoping to convince the public that even though they didn't actually *photograph* living fairies, they had *seen* them, elevates their quaint little hoax from the realm of the charming to the truly absurd. That serene stubbornness in the face of all evidence to the contrary is a trait Elsie and Frances shared with the best of the legendary liars.

Speaking of liars, consider two men, perpetrators of the "de Loys' ape" hoax. You can't truly appreciate this one until you see the "single surviving picture" of the creature, a "tailless primate" discovered in the uncharted jungles of South America near the border between Columbia and Venezuela. *Ameranthropoides loysi* was the scientific name suggested by Swiss geological explorer François de Loys. The animal was promoted by de Loys' fellow explorer, George Montandon, as a previously unknown species. The men reported they had shot and killed it when, along with its mate, the apes began defecating into their hands and throwing feces at the explorers. (That isn't unusual. I once witnessed an enormous silverback at the San Diego zoo neatly nail an obnoxious heckler in the face with a similar maneuver eliciting a howl of anguish from the ape's annoying tormentor. The rest of us who had seen the altercation cheered—for the gorilla, of course.)

The poor creature was just a spider monkey, a pet of de Loys, that had died. Its tail had been amputated earlier, due to an injury or illness of some sort. The macabre monkey was set on a small stool with a long stick propped against its chin holding it upright in rigor mortis, an undignified end for the unfortunate little primate. People *still* believe, based on the photo, that a species of large hominid remains hidden in South America. For purposes of deception, anything in the photo's background that would have suggested the animal was really small, not large, was carefully removed. Makes you almost feel like throwing something at the perpetrators of the hoax yourself, doesn't it?

Ever heard of the Pacific Northwest tree octopus? I hadn't, either, until quite recently. This elaborately concocted hoax is both laughable and representative of the tongue-in-cheek kind of hoax, the "let's see how far we can push this thing" sort. It's a prank amped-up college students might concoct during an all-nighter while buzzed on far too much Red Bull.

Accordingly to the website they created in support of their hoax:

> The Pacific Northwest tree octopus (*Octopus paxarbolis*) can be found in the temperate rainforests of the Olympic Peninsula on the west coast of North America. Their habitat lies on the eastern side of the Olympic mountain range, adjacent to Hood Canal. These solitary cephalopods reach an average size (measured from arm-tip to mantle-tip,) of 30-33 cm [around 12 inches]. Unlike most other cephalopods, tree octopuses are amphibious, spending only their early life and the period of their mating season in their ancestral aquatic environment. Because of the moistness of the rainforests and specialized skin adaptations, they are able to keep from becoming desiccated for prolonged periods of time, but given the chance they would prefer resting in pooled water. An intelligent and inquisitive being (it has the largest brain-to-body ratio of any mollusk), the tree octopus explores its arboreal world by both touch and sight.

The preceding information is the product of Lyle Zapato's fertile imagination. In one study, all twenty-five seventh graders in a research class recommended the site to others and basically swallowed the bogus information hook, line, and sinker. I can understand why—the deception is a spot-on parody of every other "save-the-whatever" campaign you've ever seen. In the same way you can't always tell when *Saturday Night Live* is showing a commercial or when it's actually a humorous skit, the tree octopus information is so much like the real thing it's almost eerie.

Because of this long tradition of fakes, most people's instinct today is to discount photos, grainy and smudged or otherwise, as unreliable. It's the boy-who-cried-wolf principle. We've been fooled once too often to fall for that trick again. The same goes for eyewitness accounts. When someone, anyone, rhetorically asks "why would I...fill-in-the-blank," you can lay serious odds that they are guilty of doing exactly what it is they swear they didn't do. That little phrase is more reliable than any lie-detector test on the planet. "Why would I say I saw a sea serpent in Loch Ness?" "Why would I say I saw fairies dancing in the dell?" "Why would I spend time and money coming up with a 'missing link' skull that wasn't genuine?" The fact that a person asks the telltale "why would I" question is practically proof of guilt—it's as simple as that.

Regardless of an otherwise impeccable character, just about anyone is capable of pulling a fast one. Fame, professional recognition, cash, you name it—there are as many motives as there are hoaxers. For people like you and me, those of us who are genuinely interested in and curious about amazing historical discoveries and those discoveries yet to come in the natural world, it makes the task of separating deliberate fakes from fascinating phenomena that much more difficult. Will we ever stop hoping? No, we won't—or at least I won't. Will we carefully consider and evaluate any and all evidence? Most definitely, and if anyone reports the unsubstantiated discovery of a fabulous monster followed by the telltale question, "Why would I make this up?" you know what you must do—turn and walk away. You're far too savvy to fall for their hoax.

Acknowledgments

For their love and support, thanks to my beloved family, Elizabeth and her husband, Amos, and their children, Katherine and Drew, Rob and his wife, Catherine, and their new baby son, William. My children and I miss Ron every day, but we carry his love and our memories of him with us always. My darling Irish dad, Andrew McGann, was a believer in the mysteries of the natural world and told me stories about mystical creatures all my life, tales that fascinated then and continue to enthrall me. Katherine, my sweet granddaughter, listened to each chapter as it was written and discussed the research about all the monsters with me. Her enthusiasm for this project matches my own, her bravery and bright spirit in the face of adversity continue to inspire me, and so this book is fondly dedicated to her.

Zebedia Wahls, adored former student and immensely talented artist, made my conception of the monsters come to life on the cover. This is the third book cover she has done for me, and I treasure every one. Zebby is also the artist who created The Woman Card[s], along with her older brother Zach, another favorite student who continues to excel at Princeton University in his graduate studies. Zeb is a truly amazing person and wise beyond her years.

Another phenomenal student of mine and the most intelligent person I've ever known, Charles Gordon Bourjaily, did an in-depth extended learning project about cryptids and cryptozoology when he was still in elementary school. Gordon is now a graduate of Amherst College and Harvard Law School, so clearly a brilliant intellect does not preclude an interest in "monsters among us."

My dear friends Anne, Linda, and Meg have stood right beside me through the decades, and I will never be able to thank them enough for their warm friendship and all it has meant to me. Their kindness, compassion, and affection have seen me though both the good times and the rough ones. I am incredibly fortunate to have each of them in my life which is constantly enriched by their presence. Love you, ladies. Always will.

Book Club friends Becky, Dell, Jane, Margaret, Sheral, and Wendy encourage me when I need it most, laugh with me whenever we get together, and commiserate with me when that's all that's left to do. Knowing them has been a privilege and a pleasure. Their sharp wit, intelligence, and depth of character show me how very much strong, caring women matter in the world.

Finally, I am forever grateful to the best editor a writer could ever hope to have, Bob McLain. Bob is both brilliant and hilarious, not to mention perennially pellucid and perspicacious. Most of us do well to juggle two or three balls in the air at a time; Bob keeps a hundred aloft—and that's just using one hand! His confidence in my writing opened a new chapter in my life. He's not only a true professional, but a wonderful person as well. Thanks, Bob.

About the Author

Andrea McGann Keech grew up in southern California. Thanks to the influence her Irish dad's early stories, she has nurtured a life-long interest in those strange and mysterious creatures known as cryptids.

She and husband Ron met at Occidental College and were married in San Francisco. Ron graduated from the University of California at San Francisco Medical School, and Andrea finished college at the University of San Francisco. They lived in Portland, Oregon, for six years and then moved to Iowa City where he was a professor of medicine and a surgeon in the department of ophthalmology at the University of Iowa for twenty-two years until his death. Their children Elizabeth and Robert made them very proud and very happy parents. Liz is an attorney, and Rob is a dentist.

Andrea taught students in English and Spanish in grades K-12 during her teaching career. She was a member of the National Assessment of Educational Progress Committee that established Writing Standards, 2011–2018, for students in grades 3–12. She has written for a variety of national educational journals and presented often at teaching conferences, but the most fulfilling aspect of her work, by far, was seeing her students succeed. Andrea lives in Iowa City with Shadow and Sunny, two wild and crazy standard poodles. Her most fulfilling role is that of playing Mary Poppins to beloved grandchildren Katherine and Drew and spending time with baby grandson William.

Her other books include *The Cream of the Crop: Tour Guide Tales from Disneyland's Golden Years, The Indulgent Grandparent's Guide to Walt Disney World,* and *Treasure of the Ten Tags, a Disneyland Adventure,* all published by Theme Park Press.